Mechanised
Livestock Feeding

Mechanised Livestock Feeding

David L. Bebb

BSP PROFESSIONAL BOOKS

OXFORD LONDON EDINBURGH

BOSTON MELBOURNE

First published 1990

British Library
Cataloguing in Publication Data
Bebb, David L.
  Mechanised livestock feeding.
  1. Livestock: feeding
  I. Title
  636.08′4

ISBN 0-632-02035-0

BSP Professional Books
A division of Blackwell Scientific
  Publications Ltd
Editorial Offices:
Osney Mead, Oxford OX2 0EL
  (Orders: Tel. 0865 240201)
25 John Street, London WC1N 2BL
23 Ainslie Place, Edinburgh EH3 6AJ
3 Cambridge Center, Suite 208, Cambridge
  MA 02142, USA
54 University Street, Carlton,
  Victoria 3053, Australia

Set by Best-set Typesetter Ltd, Hong Kong
Printed and bound in Great Britain by
Cambridge University Press, Cambridge

Contents

feeding systems – Floor/slat feeding systems – Centreless auger
feed distribution system – Cable and disc feed distribution
system – Mechanised feeding of caged birds

Preface

Feeding is a key operation on intensive livestock farms where it is not only the quantity and quality of feed – but also the frequency and especially accuracy with which it is dispensed, which may have a particularly significant effect upon enterprise profitability. Thus, for instance, poor design of milking parlour feed dispensers and lack of calibration and maintenance can easily lose some producers £50–£80/cow annually. For the stockperson, feeding can be both time consuming and tedious: mechanising the job can release time which can be more profitably devoted towards enhanced stockmanship and enterprise control. Here opportunities are created by new technology, often incorporating the ubiquitous microchip, for tighter management and improved feedback on livestock performance: examples include electronic weighing and logging of feed, and individual electronic feeding and monitoring of dairy cows, calves, and sows. Appropriate choice of equipment can create opportunity for entrepreneurial buying of alternative feeds, including food by-products – the practice of which on some livestock units can, for instance, often make the difference between profit and loss in the volatile pig market.

There is a considerable annual investment in livestock feeding equipment – yet there is little comprehensive written guidance available for the farmer. The financial penalty for a wrong decision is profound – and there are many instances where pertinent factors have not always been adequately considered before costly decisions have been made. This book sets out to provide a comprehensive guide to the farmer, adviser, supplier, and student, to enable them to appreciate both the biological and engineering factors affecting choice of equipment – and also how to get the best out of it.

David Bebb

Acknowledgements

The Author gratefully acknowledges his indebtedness to Vivian Baker for typing the manuscript and Stefan Oakley and Andrew Haslock for assistance with drawings and photographs.

Thanks are expressed to the following companies/organisations for the supply of illustration material:

Alfa Laval, Peter Allen Milking Machines, Alvan Blanch, ATL Agricultural Technology Limited, Audureau Farm Machinery, AZA, Bentall Simplex, Ben Burgess, Big Dutchman, BOM, Boythorpe, Chilton, Chore Time Elite NV, Christy Hunt Agricultural, Ernest Collinson Ltd, Cormall, DOL, EB Equipment Limited, Ferrag Gehl, Fullwood, Funki, Hampshire Feeding Systems Limited, Hareland Engineering, Hartlink, Hyde Hall (Rettendon) Ltd, Richard Keenan Ltd, Kidd Farm Machinery, Kongskilde, Law Denis, Lucas, Manus, Montec Systems, Multimate, Newlands, Orby Engineering Limited, P J Parmiter, Permastore, Philco Dierings, Rowlands Silage and Storage Systems, Taylor Fosse, Waterson Engineering, Richard Western Limited, Westfalia Separator, Vicon, Volac.

Acknowledgement is made to the following organisations for permission to use copyright material specified in text:

Centre for Rural Building, Aberdeen
Dairy Farmer, Ipswich
Electricity Council, Stoneleigh
Ministry of Agriculture, Fisheries and Food, London
Scottish Agricultural Colleges, Ayr

1 On-the-farm concentrate feed preparation

Introduction

Ever tightening margins besetting intensive livestock units prompt a closer look at the feasibility of producing compounds on the farm. With concentrate feed consuming from 35% of the gross output for dairy cattle to 80% for pigs and poultry, there is plenty of scope to make very significant savings. It is difficult to generalise, and in establishing the economic benefits (if any) of home mill and mixing, much depends on the individual circumstances of farm and farmer. Capital investment and depreciation, together with annual tonnage, are the most dominant factors affecting potential savings. Such savings, however, can be quickly eroded if feed quality is impaired, and indeed, recent Meat and Livestock Commission (MLC) evidence indicates that, typically, home mixers feeding fattening pigs achieve a typical 0.23 feed conversion ratio (FCR) shortfall (equivalent to approximately £1.75 per pig from 30 to 80 kg) compared to those bought in compound users. In some cases, the consequences of overall inferior FCR figures can wipe out the savings made. Nevertheless, a combination of factors – not least the opportunity of avoiding co-responsibility levy by cereal growers with intensive livestock units – is prompting many producers to consider some alternative to bought-in compounds.

Systems appraisal

The unique circumstances of each livestock enterprise prompts the need to appraise carefully the specific benefits and disadvantages of the alternative means of procuring feedstuffs for that unit.

Purchasing compounds

The protagonist of using compounds will quickly point out the advantage of a guaranteed nutritive analysis, but there is no ingredient specification, although this may soon be made a legal requirement. There is much debate at present within the agricultural industry on this point. Some producers are critical of the use of non-cereal 'fillers' by compounders, which contribute nothing towards the European grain mountain. However, the compounder may have better knowledge of the ingredients' market and, perhaps, may be able to purchase more advantageously than home mixers through using forward contracts and the futures market. Indeed, following the introduction of milk quotas, the compound industry has now entered a particularly competitive era, and home producers would be well advised to seek a series of regular quotations for compounds, exploiting bulk discounts, and comparing them to the true costs of home mill and mix rations. Remember that the compounder may offer not only discounts but also credit facilities and delivery in bulk to bin storage within the livestock enterprise minimises on-farm transport difficulties and investment.

Perhaps the major factor in favour of compounds is that potential capital investment is reduced. It should also be pointed out that a change to home mill and mixing can have a dramatic effect on the business's cash flow. This might be most marked, perhaps, with the farmer who is used to 30 or 60 days' credit with his feed merchant and sells his grain immediately after harvest. He would be well advised to warn the bank manager!

Feeding compounds also entail a lower requirement in respect to labour and management supervision, with less worries over potential mechanical breakdowns.

Home mill and mixing

The dominant factor in favour of home mill and mixing, of course, is the potential to make significant savings over compounds. Gross savings achieved vary considerably from enterprise to enterprise but £20 to £30 per tonne is not uncommon. Typically £6 to £15 per tonne must be deducted for home milling costs, but nevertheless net savings achieved are often very worthwhile and, indeed, some producers would argue that in depressed markets they are the key to sustaining a viable enterprise.

Thus, if a net saving of £15 is taken, and if there is no detriment to FCRs, the overall annual savings can amount to over £10000 for a 100 sow unit with progeny on to bacon, £1500 to £1800 for a 100 cow dairy unit, and £3500 for a 5000 laying bird poultry unit.

Fundamentally, home mill and mixing does not have the cost of sales and administration, transport and distribution, or profit required by a compounder. A large compounder recently attributed 80% of the cost of a compound to raw materials and 20% to the afore listed items.

Of course, the compounder is better placed to buy ingredients more cheaply. That said, there are growing examples of producers who, with the aid of professional advisers, are able, too, to play the ingredients' market by 'opportunity buying'. Thus, the price of soya can vary by as much as £50 per tonne during any 12-month period. Livestock producers' entrepreneurial skills can develop into the most decisive factor governing profitability of a livestock enterprise.

Home mill and mix can be best exploited where cereals are home grown. A reliable outlet for grain is thus assured, and the farmer is divested of the lottery of when to dispose of it, as well as saving on the costs of transport. Furthermore, costs of grain cleaning and drying may, in some circumstances, be reduced – as well as a ready market being available for grain of questionable quality (which should, of course, be taken into consideration in compiling rations). The existence of grain handling and storage facilities does encourage the potential to incorporate a mill and mix plant into or around them, with important economies in costs of buildings, conveying and storage.

After potential saving, many home mixers like the freedom and flexibility of control associated with the system. The exact ingredients are known and fresh rations, minerals and growth promoters are always available. Furthermore, medicinal additives can be brought into use within a very short timespan. However, recent legislation now places strict control on the incorporation of such additives where the home mixer using them is required to register and be subject to scrutiny.

On-farm contract mixing

On-farm contract mill and mix is an ideal means of 'dipping a toe in the water' with minimum outlay and risk. A number of firms, such as Mill Feed Services, offer a comprehensive service whereby bought-in or home-grown cereals are milled and mixed with purchased protein additives, with the opportunity of incorporating chopped straw if need be. Furthermore, round-the-farm transport of mixed concentrates is a further potential incentive. Many of the supposed advantages of home mixing are also available with this method; for example, fresh rations and existing grain handling and storage systems can be utilised. On-farm contract mixers also claim that rations of more consistent quality and assured nutritional value are possible than with traditional home-mixed rations.

The major attraction, however, is in terms of elimination of capital investment in plant and significantly reduced labour input. Charges vary and depend upon frequency, quantities, distance and fineness of grinding. With less than 150 tonnes per annum, the higher overhead depreciation charges for buying one's own plant can easily mean that the contract service is more cost effective, as has been proven in various recent surveys. Once again, it pays to obtain quotes and cost out carefully!

Co-operation

Where producers are prepared to sacrifice some degree of their 'independence', individual capital investment can be significantly reduced and labour input also trimmed by a co-operative mill and mix unit. There are obvious opportunities to spread interest, depreciation and labour charges over larger tonnage outputs, and so possibly achieve lower ration costs than for smaller independent units. These savings will be partially offset by greater transport costs. This type of feed processing appears to have been most successful where a group of farmers are looking to use broadly similar rations, for example, a group of bacon producers.

It is likely that particular attention will be paid to procurement of raw materials and quality control in the production of rations, which may enhance performance compared to smaller mill and plant units.

Where such a co-operative venture is contemplated, it may be more appropriate to make use of robust secondhand 'industrial' quality equipment from old feed mills, which is capable of high outputs and extended running time. In such circumstances, the continued availability of secondhand parts is a vital factor.

Factors to consider when planning a home mill and mix unit

When appraising the considerations for a potential mill and mix plant, the selection of appropriate hardware is perhaps the easiest part. There are equally important factors which must be considered first. For convenience these contributory aspects can be considered under the headings of:

(1) Stock requirements.
(2) Operational considerations.
(3) Cost of feed.
(4) Buildings, electrical supply and equipment.

(1) Stock requirements

Here the type and numbers of livestock, and quantities and form of final feed are primary considerations. Table 1.1 can be used as a guide in determining annual requirements.

A low cost plant, based on a roller mill and mixer producing rations suitable for dairy or beef cattle, is normally justified where there is an annual consumption of not less than 100 tonnes. Pigs and poultry require a finer grist, necessitating a hammer mill, and as an aid to palatability and means of reducing waste a cuber or pelleter may be desirable. However, cubers are expensive and outputs are relatively low. Additional storage and cooling bins are needed. Such higher investment and operating costs require a higher annual throughput to break even, the point of which is influenced by depreciation charges.

The method of feeding will to some extent influence what is required within the mill and mix plant. Thus, for floor feeding of pigs cubing becomes particularly worthwhile, whereas it can be avoided when liquid feeding.

The number, type and complexity of diets is also a crucial factor. Small batches of many different rations create operational complexities and often it is not really feasible to produce highly specialised rations, such as creep feed pellets. It is vitally important that the mill and mix system should be capable of being easily set up for producing different rations with the minimum of fuss, and that such rations can be repeated accurately.

Table 1.1 Typical annual feeding requirements for different numbers and classes of stock.

Stock	Tonnes per annum	Typical peak weekly requirement (tonnes)
Pigs		
100 sows	120	2.5
Annual progeny of 100 sows (2200 pigs) to:		
pork weight (67 kg)	350	6.8
cutter weight (88 kg)	530	10.2
bacon weight (91 kg)	550	10.6
Dairy		
100 dairy cows being fed 1 tonne/year	100	6.0
Beef		
100 cereal beef	230	6.1
100 18-month grass/cereal beef	60	2.1
Poultry		
5000 laying hens fed 125 g/day	230	4.5
Broiler unit producing 100 000 birds per annum to 2.1 kg	452	2.2

(2) Operational considerations

Continued successful operation of on-the-farm mill and mix plant is dependent upon adequate availability of labour, the ability to seek supplies of appropriate raw materials at the right price, taking account of milling losses and dust, and maintaining high standards of quality and accuracy.

It is fundamental that adequate labour is available to look after the mill and mix plant – adequate in terms of hours available and quality of manpower. It is obviously harder to integrate a proposed mill and mix plant onto an already fully deployed workforce. Taking on extra staff may not be justified. Equally, stock persons are not renowned for their mechanical ability or aptitude, so it is important to make the right choice in the person to put in charge of operations. Here labour input is more than merely setting up and supervising the plant and handling materials. Regular on-going managerial time input has to be found for purchasing ingredients and formulating rations. Enthusiasm is required to keep in touch with market trends and prices, and the farmer/manager must be prepared to spend much time on the telephone maintaining contact with a number of merchants. In short, the team, whether it is one individual or a combination of stockman and manager, will find itself having to learn a manufacturing skill and also quickly acquiring a knack for buying raw materials.

The degree of automation employed may ease labour supervision and give potential for improving accuracy in production of rations. As a rough guide, labour requirements range from about 0.75 man hours per tonne mixed to about 0.25 hours per tonne on a reasonably automated unit.

Quality control is a cardinal rule in on-the-farm production of rations. It is fatuous saving £20 per tonne on feed costs if performance is depressed by £25 per tonne fed. In circumstances where the margin of savings from home milling and mixing is small in comparison to purchasing compounds, a small reduction in feed conversion efficiency could significantly reduce the balance in favour of home mixing or indeed produce a loss situation. Quality control effectively means having responsible, trained and intelligent staff, ensuring accuracy in the metering or weighing of raw materials, care and maintenance of equipment, sound feed formulation and nutritional advice, and not least arranging for regular laboratory analyses of ingredients (especially the variable ones). There is considerable scope for wider uptake of feed analysis services with potential benefits of avoiding underperformance in FCRs. Most members of the mineral/vitamin supplement supply trade will provide on-going food analysis on behalf of their customers, but it is important that enough samples are taken and that the results are available and acted upon with the minimum of delay.

Pneumatic conveying of milled material normally requires the inclusion of a cyclone and dust socks. Inevitably such conveying methods lead to grinding losses of around 1% or so, or even higher in some circumstances, and in addition to the losses themselves, there is the added adverse effect on the working environment. Legal requirements now entail assessment of the extent of the dust problem and minimising of the source. The installation of extractor fans is not truly an adequate answer and care needs to be taken to minimise fire risks. A vacuum cleaner is recommended for keeping the installation clean to minimise dust and attack from vermin.

(3) Cost of feed preparation

Some attempt should always be made to estimate the costs per tonne of prepared feed and relate this to the annual amount prepared on the farm. The detailed costing will vary depending on individual circumstances.

In assessing variable costs, labour forms the biggest component; but for some smaller units where no extra labour is taken on, in reality no labour cost is involved. The adoption of automatic controls, including computer control, will help to reduce labour costs. Electricity is the next largest variable cost, involving from 14 to 21 kWh per tonne mixed. Automatic controls can facilitate the possibility of using Economy 7 tariffs thus significantly reducing costs, but operational noise through the night may pose problems in respect to nearby houses.

Whereas the variable costs are fairly easy to calculate, in deriving fixed costs some will borrow, others will self finance and forego the interest which might have accrued by investing the capital in a different way. Consideration of building costs will also have a marked effect, since adaptation of old existing buildings will provide a major economy compared to investment in new buildings. Do not underestimate what is required. It should be remembered that bin storage and conveyors will make up at least half of the capital investment in equipment.

In the following examples it is assumed that interest has been charged at 15%, payable on the reducing balance as a medium term loan over 5 years, with repayments made in ten equally spaced instalments over the 5-year period. On this basis the total interest charged for each £100 borrowed would be £41.25. It is not unreasonable to write off the investment and interest over 10 years giving an annual interest and depreciation charge of £14.13 per annum for each £100 capital. A shorter depreciation period would obviously increase this charge by a proportionate amount.

In these costings it is assumed that management and incidental costs in purchasing feed have been included in the 'true cost' of materials bought in. Also, no attempt has been made to calculate an interest charge for borrowing for forward buying as this depends on individual circumstances.

Electricity has been costed at an average unit charge of 6.1p. However, the actual cost will be a function of which tariff the farm now adopts, be it Small Supplies General, Economy 7, or Evening/Weekend Economy 7 tariff, Seasonal Time of Day tariff, or Maximum Demand tariff. The costings also ignore any possible effect of impaired stock FCRs of home produced rations compared to bought-in compounds. Any difference can be minimised by close attention to ration formulation and regular analysis of raw materials. Thus, £1/tonne for feed analysis has been included.

In Examples 2 and 3 the installation of a cuber/pelleter would have a significant effect. Investment of an additional minimum of approximately £5000 (as appropriate for Example 2) would be necessary, not only for the cuber/pelleter, but also for cooling bins and bucket elevator. The specific output of cubers is relatively low and, with cooling also taken into consideration, an additional £1.20–3.00 per tonne should be allocated for electricity, depending on the size of cube/pellet produced. Depreciation and interest charges must be added, of £2.82 or £5.65 per tonne (10 and 5 year depreciation respectively) for 250 tonnes per annum, and £1.41 or £2.83 per tonne (10 and 5 year depreciation respectively) for 500 tonnes per annum. Overall maintenance and repair and, to some extent, labour costs will also be increased. Thus, cubing/pelleting is likely to add *at least* £4.00 per tonne and £5.50–6.00 is more probable.

A further cost to be taken into account, and often ignored, is the transport cost from mill and mix plant to the livestock units. Typically this is built into the cost of supply of compound feeds, or in the charges made by a contract mobile mill and mix service. This charge can only be calculated in relation to the specific farm circumstances. Typically, dry feed is transported round the farm by bulk feed trailer, conveying system or 0.5–1.00 tonne tote bins transported by forklift.

On the farm feed processing costs are thus highly geared to annual throughputs and the length of chosen depreciation period. As depreciation and interest on capital consume a major component of total costs, particularly at lower outputs, it may be possible to spread these overheads in a co-operative enterprise.

Example 1   System for dairy and/or cattle, typical throughput of 100–250 tonnes per annum.

Capital costs		£
£10 000 in total for		
2.25 kW roller mill		1 300
1 tonne chain and slat mixer		2 300
1 weigh hopper with dial weigher		1 200
2 augers		800
2 × 20 tonne raw material bins		4 400

	Tonnes produced annually	
	100	200
	Cost per tonne produced	
	£	£
Depreciation and interest charges*		
(£10 000 × £14.13 per £100		
capital ÷ annual tonnage)	14.13	7.07
Repairs and maintenance	0.50	0.50
Electricity: 14 kWh per tonne at		
6.1p per kWh	0.85	0.85
Milling losses (0.5% of 800 kg		
at £110 per tonne)	0.44	0.44
Allocation for labour/management		
costs at 0.5 person hour per tonne	2.00	2.00
Charge for raw material analysis	1.00	1.00
Total cost per tonne	£18.92	£11.86

* If the equipment is written off over 5 years instead of 10, the relevant interest/depreciation figures are doubled and the total cost per tonne becomes £33.05 and £18.93 respectively

It should be remembered that raw material storage bins consume 44% of equipment costs. If existing grain storage bins can be utilised this can make a considerable difference to overall depreciation costs.

Example 2 System for pigs (sows plus progeny), typical throughput 250–500 tonnes per annum.

	Capital costs	£
	£8 800 total equipment cost for combined 5.5 kW hammer mill with nominal 1.75–2 tonne measuring hopper and nominal 0.9–1.0 tonne vertical mixer unit	5 000
	1 only 20 tonne grain bin	2 200
	2 auger conveyors	800
	1 mixed feed storage bin	800

excluding electrical installation and building costs

	Tonnes produced annually	
	250	500
	Cost per tonne produced	
	£	£
Depreciation and interest charges* (£8 800 × £14.13 per £100 capital ÷ annual tonnage)	4.97	2.49
Repairs and maintenance	0.60	0.60
Electricity: 20 kWh per tonne at 6.1p per kWh	1.22	1.22
Milling losses (1.5% of 800 kg at £110 per tonne)	1.32	1.32
Allocation for labour/management costs at 0.5 person hours per tonne	2.00	2.00
Charge for raw material analysis	1.00	1.00
Total cost per tonne	£11.11	£8.63

* If depreciation is taken over 5 years rather than 10 the total cost per tonne becomes £16.08 and £11.12 respectively

Example 3 System for pigs (sows plus progeny), typical annual throughput 1200–1800 tonnes.

Capital costs

	£
£23 000 in total equipment cost for	
11 kW hammer mill with 2	
tonne vertical mixer unit	
mounted on load cells with	
auto control	8 000
6 augers	2 400
5 raw material bins	10 000
3 mixed material bins	2 600

excluding any electrical sub main and building costs

	Tonnes produced annually	
	1 200	1 800
	Cost per tonne produced	
	£	£
Depreciation and interest charges* (£23 800 × £14.13 per £100 capital ÷ annual tonnage)	2.71	1.81
Repairs and maintenance	0.60	0.60
Electricity: 21 kWh per tonne at 6.1p per kWh	1.28	1.28
Milling losses (1.5% of 800 kg at £110 per tonne)	1.32	1.32
Allocation for labour/management costs at 0.25 person hours per tonne	1.00	1.00
Total cost per tonne	£8.91	£8.01

* If depreciation is taken over 5 years not 10, the total cost per tonne becomes £11.62 and £9.82 respectively

Example 4 Mobile mill and mix system for cattle or pigs, typical throughput 300–800 tonnes per annum.

Capital costs
£13 000 for tractor drawn and
driven mill mix system with
electronic weighing requiring 75
kW tractor to operate it, and
with output of 2 tonnes per hour,
excluding cost of ingredient
storage and capital cost of tractor

	Tonnes produced annually	
	300	800
	Cost per tonne produced	
	£	£
Depreciation and interest charges* (5-year write-off, with capital cost £13 000, less £2000 trade-in at end of 5 years, net cost £11 000 × £28.25 per £100 capital ÷ annual tonnage)	10.36	3.88
Labour and tractor costs[†] (estimated cost per hour £17.00, thus cost per tonne at output of 2 tonne per hour)	8.50	8.50
Maintenance and repair	1.00	1.00
Milling losses (1% of 800 kg at £110 per tonne)	0.80	0.80
Charge for raw material analysis	1.00	1.00
Total cost per tonne	£21.74	£15.26

* With a throughput of 300 tonnes per annum it may be feasible to consider write-off over more than 5 years, say 7 years, which would reduce the depreciation cost for 300 tonnes per annum by £2.96
[†] The full costs of labour and tractor have been included. In practice, some would argue that a lesser charge may be appropriate as the tractor may need to be owned anyway for other farm work

Buildings, electrical supply and equipment: fixed and mobile plant

BUILDINGS

The availability of existing farm buildings suitable for installation of mill and mix plant can often tip the balance in favour of on-farm feed processing. But existing buildings are often beset with several constraints on the designer – not least height limitations and width and height of doorways. The building should be as high as possible to accommodate tall bins and to make use of gravity for as much conveying as possible. It is advisable to provide sufficient floor space not only for appropriate equipment access and servicing, but also for floor storage of bags of concentrates and to facilitate future installation of alternative equipment. In many respects it may be desirable to position equipment so that there is a straight-line flow of in-coming ingredients at one end of the building and out-going rations at the other. It is essential to make adequate space allowance for access for deliveries and collections, including ensuring that doors are large enough to allow access for the farm materials handler, which should be of the mechanised form rather than relying on pure human effort!

Siting is highly critical: factors which must be considered include position of existing buildings, ease of access for road transport and supply of electricity. The ideal building would probably adjoin the grain store at one end and the livestock house at the other to minimise transport of materials. As such a site is normally unlikely it is usual to incorporate the plant adjacent to the grain store, for instance as a lean-to extension. Alternatively, there may be enough floor space within the store itself, preferably within a partitioned off area, as dust and discarded ingredient bags can increase the likelihood of insect infestation in the grain store.

Plenty of roof lights (around 15% of the plan area) are highly desirable, together with a positive means for dust collection. The latter is vital, since dust is not only a hazard for work persons and stock, but also constitutes a fire risk. Extractor fans can help, but an industrial vacuum cleaner and face mask or aspirated helmet are essential. Cross-contamination really becomes a problem when drugs or growth promoters are used, and therefore such additives are best added directly into the mixer.

ELECTRICITY

The electricity supply should be adequate as considerable power is needed for mills, mixers and conveying equipment. A three-phase supply is preferable and essential for all but smaller installations. The producer contemplating installation of on-the-farm mill and mix plant must be prepared to face the necessity of the cost of installation of a higher capacity submains. Some economies in running costs can be made if Economy 7 off-peak electrical supply is used, but the savings need to be weighed against the cost of appropriate controls to allow for unattended operation during unsocial hours. Disturbance from noise at night may also need to be considered in some localities.

EQUIPMENT

Choice of system
Before consideration of individual pieces of equipment, one needs to consider the system most appropriate to the farm's circumstances. The choices can be summarised as shown in the following flowchart.

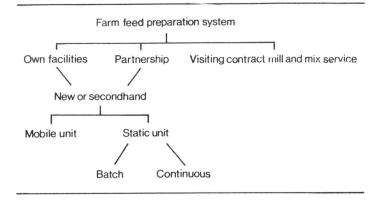

Though a visiting contract mill and mix service affords a number of obvious benefits as previously discussed, ownership of plant does give greater independence and is certainly more appropriate for larger annual throughputs. But some enlightened neighbouring producers have seized the opportunity to form partnerships which provide further potential economies and which, despite increased transport costs, can certainly dramatically reduce depreciation costs. Further, such high output plants can streamline management and enhance bulk buying discounts. Shared plant, too, can often enhance the justification of investments in high cost/low output equipment, such as a cuber or pelleter, which might be ruled out otherwise.

The normal temptation is for investment in new plant and this might be wise for the majority. However, at least one farmer group in a large co-operative venture bought up secondhand industrial plant which is more sturdy than farm mill and mix equipment and has a longer potential life span. Reputation, reliability and availability of spares for the years ahead are the most vital questions to pursue when buying secondhand.

Mobile mill and mix units
Most on-the-farm mill and mix plants are of the fixed type, but there has been a growing market in recent years for tractor-driven mobile mill mixers. Such mobile mill mixers consist of a tractor pto driven hammer mill and vertical mixer mounted on a pneumatically tyred chassis. They are capable of high outputs and larger models have a typical capacity to deal with up to 40 tonnes/week. Mixing can be done on-the-move and the unit taken anywhere to stock where, using the long hydraulically driven discharge auger, the machine thus doubles as a bulk feed transporter.

In many circumstances, the overall plant capital investment require-

Plate 1.1 Mobile mill and mix unit (*Ferrag-Gehl Mix-All*). Such units have up to 4.0 m³ mixers with capacity, it is claimed, for up to a 1500 kg mix, based on barley at 14% moisture content, to be milled, mixed and discharged in around 30 minutes. Milling is achieved using a 535 mm wide 66 hammer mill.

ment for such a system can appear quite attractive. Other factors in favour of mobile feed processing plant include the significantly reduced requirement for building space, together with no limitations on electrical supplies. However, these advantages must be balanced by the necessity of tying up a tractor of suitable pto power on a regular basis, and this must feature in the overall costing. Furthermore, by comparison to fixed plant, which can be run automatically by day or night using electricity, the overall potential labour requirement and running costs are substantial, often at least as much as the depreciation of the machine itself. It is certainly worthwhile contemplating a mobile mix system where there is a considerable arable unit as well as stock. The mill mix unit would allow extra use to be made of a large cultivation tractor and provide employment for the tractor driver in slacker months of the season.

Arrangements must be made for the handling of cereal and concentrate ingredients at a central point to facilitate rapid filling of the machine.

Fixed mill and mix plant

Static mill and mix units are designed to be run on either a batch basis or by means of a continuous flow proportioning system. For throughputs up

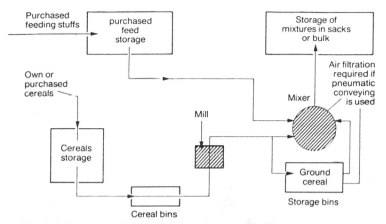

Fig. 1.1 Example of flow diagram for batch mill and mix plant.

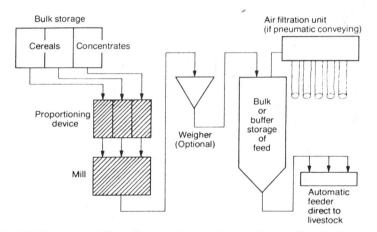

Fig. 1.2 Example of flow diagram for continuous flow mill and mix plant.

to around 600 tonnes/annum the batch system is undoubtedly the most appropriate. Either batch or continuous flow systems may be suitable for throughputs in the range 600–2000 tonnes/annum. Continuous flow systems are a more appropriate choice above 1200–1500 tonnes/annum. Here all constituents are metered simultaneously through parallel proportioning mechanisms (typically augers). Material is then delivered to the mill which functions to disintegrate those parts of the ration requiring it, and in the process all elements are thoroughly mixed. The attraction of a continuous flow system is that it lends itself to computerisation, takes up less space and ready-prepared feed can be delivered directly to buffer storage bins close to livestock housing and then to livestock by means of automatic feeding equipment. Management supervision and labour in handling are thus reduced to a minimum. However, most continuous flow systems incorporate volumetric proportioning of constituents and close attention has to be given to calibration and adjustments in order to accommodate changes in physical characteristics between batches of in-

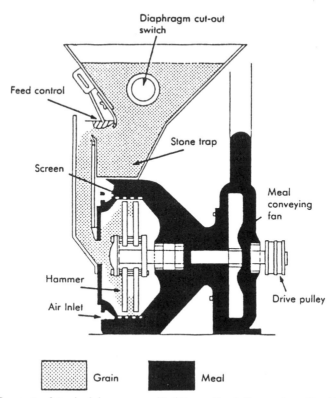

Diaphragm cut-out switch

Feed control

Stone trap

Screen

Meal conveying fan

Hammer

Air Inlet

Drive pulley

Grain   Meal

Fig. 1.3 Layout of typical hammer mill (*Farm Feed Processing*, Booklet 2125, MAFF, © Crown Copyright 1989).

gredients and thus ensure that rations are adequately balanced. It has been argued that, in some circumstances, the greater limits of measurement tolerated by such systems can reflect in inferior performance of livestock, narrowing the potential cost saving margin between home-mixed and bought-in rations.

Grinding/bruising

With the exception of sheep, which are capable of utilising whole grains, all classes of livestock require some processing of grain. For horses and ruminants all that is necessary is to bruise and crack the hard outer layer of grain, using a roller or bruising mill, to produce a flaked material. The objective is to slow down the rate of digestion and give a longer rumen retention time. For effective digestion in pigs and poultry it is necessary to grind material to a relatively fine grist, the degree of which is influenced by the type and age of stock, using a hammer or plate mill. Grinding should always be severe enough to eliminate whole or near whole size grains which are unlikely to be utilised efficiently by animals, and the grist should have a distinctly gritty feel, especially for poultry. A screen size of 3 mm, for example, will ensure that the largest size of grist is no more than 3 mm and that there is not excessive dust. Finer grinding not only promotes more dust but consumes extra power with no enhancement of animal performance. Indeed very fine grinding may result even in lower feed conversion in some cases.

Hammer mills

Hammer mills comprise free swinging hammers rotating at speeds up to 6000 rpm which smash cereal grains introduced into their path until the grist formed is small enough to escape through the holes of a replaceable screen which surrounds the whole or part of the hammer chamber. Usually, hammers are mounted on a shaft in the horizontal axis, but some are designed to rotate in the vertical axis. The latter are claimed to be effective in grinding grain of high moisture content, that is up to 24%.

Grist is removed from the mill either by a self-contained pneumatic conveying system delivering meal by ductwork to a mixer or storage or with a separately driven auger conveying meal from the base of the mill. Pneumatic conveying, using a fan integrally fixed to the common drive shaft, is perhaps the most convenient as the duct work can readily negotiate a complex route, including a number of bends, to the desired destination which could be even up to 60 metres away. Unless air is directly recycled

Plate 1.2 Hammer mill (*Christy Hunt Agricultural Ltd*).

Plate 1.3 Hammer mill with cyclone and filter socks (*Alvan Blanch*).

back to the mill from the mixer, provision has to be made at the point of delivery to separate the meal from the air and allow the air to escape. This is achieved using a cyclone, a conical shaped container with its point at the base, and filter socks.

The cyclone should be designed to match the size of mill, but it can often be dispensed with where a vertical mixer is used, the shape of which doubles as a large cyclone. Filter socks need to provide around 1.5 m² of sock area for each kW mill size and should be of appropriate weave density to trap dust and yet release air. Instead of blowing milled grist, some hammer mills are designed for mounting directly on top of the mill with pneumatic delivery of unmilled grain up a suction pipe, obviating the need for a separate grain conveyor. The main problem with pneumatic conveying is that moving air is costly in power consumed. The fan secondary air intake shutter must be set to ensure that power is not wasted and that just sufficient air is allowed to convey the feed without blocking, ensuring that as much power as possible is available to drive the hammers. By comparison, hammer mills with mechanical means of removing meal from behind the screen have a higher specific output (increased output for the same power input). Table 1.2 demonstrates this point and shows comparative typical outputs for mills of various size. Hammer mills are supplied with a range of screens with various mesh to produce rations to suit different stock. Typical screen sizes are shown in Table 1.3. Screen

Table 1.2 Approximate output in kg/h for different sizes and types of hammer mill.

Screen size	Mill size (kW)						Conveying method
	3.8		5.6		7.5		
mm	3.00	6.5	3.00	6.5	3.00	6.5	
(in)	($\frac{1}{8}$)	($\frac{1}{4}$)	($\frac{1}{8}$)	($\frac{1}{4}$)	($\frac{1}{8}$)	($\frac{1}{4}$)	
Barley	250	500	400	750	525	1000	Pneumatic
	300	800	450	1200	600	1600	Mechanical
Wheat	325	600	550	850	650	1150	Pneumatic
	350	1000	575	1500	700	2000	Mechanical

Table 1.3 Screen sizes for different classes of stock.

	Screen size (mm)
All cattle	6.5
Sheep	3.0–12.5
Fattening pigs*	2.5–4.0
Young pigs	1.5–3.0
Poultry	3.0–4.0

* where pigs are pipeline fed on liquid/ meal mixtures, maximum screen size should be 3 mm to reduce the possibility of pipeline blockages

Table 1.4 Typical output in kg/h from a 3.7 kW mill.

Crop	Coarse	Medium	Fine
Oats	350	150	75
Wheat, maize, beans	350	200	125
Barley	350	200	100

size, type of cereal, and moisture content all have significant effects on output, as illustrated in Tables 1.4 and 1.5.

To operate the mill, it should be allowed to gain full working speed before feeding any grain to the hammers. With the conventional gravity

Table 1.5 Variation in mill output with grain
moisture content.

Moisture content	Specific output (kg meal per kWh electrical energy consumed)
13	54.4
15	47.2
17	40.6
19	34.0

feed hammer mill there are two basic settings. There is an adjustable feed
gate to supply grain to the intake hopper, and also an adjustable flap,
sometimes counterweight operated, which controls the amount of air
drawn directly down the feed tube or through the grain from the hopper.
If the flap is set to throttle the amount of air taken in directly at the top
of the vertical feed tube then more air must be drawn down instead
through the grain from the intake hopper. This encourages a greater flow
of grain to the hammers. Usually a built-in ammeter helps in setting an
appropriate flow rate to ensure that the mill will not stall but is working
at its correct power rating.

It is advisable to select a hammer mill with an easily emptied stone/
metal trap to minimise damage to the screens (as well as to minimise the
dangers to livestock). It is also desirable to fit an integral diaphragm switch
on the side of the intake hopper to shut down the mill automatically when
the grain supply has run out. Other features should include a large screen
area with means of achieving even flow of grain to it for maximum output.
There should be easy access to change screens, and hammers should be
reversible to provide a fresh striking edge. On most mills all four corners of
the hammers can be used for reversing them and turning them end to end.
Worn hammers, of course, reduce the specific output and they should be
frequently inspected and always turned and replaced as a set. Any broken/
damaged hammer causes considerable imbalance, vibration and bearing
wear: replacement of the set is the only answer. When checking hammers
it is advisable, also, to inspect screens for signs of wear and any damage
caused by foreign objects. Periodic checks need to be made, too, on belt
tension to ensure that full power is transmitted.

Plate mills

Plate mills, though not now commonly used, have a number of particular
advantages. They can grind grain of higher moisture contents (even up to

30% is claimed) and they are capable of producing a wide range of grist, from a product not too dissimilar from that produced by hammer mills to a 'kibbled' product where grain is broken into coarse fragments. But they are not particularly appropriate for fine grinding. By comparison with the hammer mill they have a good specific output of around 70 kg/kWh for meal grinding, to about double this for kibbling.

The plate mill comprises two cast iron dished plates with teeth or serrations on one or both faces. They are mounted on a common shaft with one plate driven, the other stationary, and pressed together by a clamping screw acting on one of the plates through a strong spring. A smaller spiral spring is fitted between the plates which prevents them touching when no grain is being fed to the mill. Grain passes over a small oscillating screen which removes large foreign objects, and is then fed to the centre of the plates when it passes towards the periphery, being gradually ground in the process. Fineness of grinding depends on the nature of the corrugations and the clearance between the plates. Plate mills should never be run empty and an automatic cut-off must be in operation before the grain feed has run out.

Plate 1.4 Plate mill (*Cormall*).

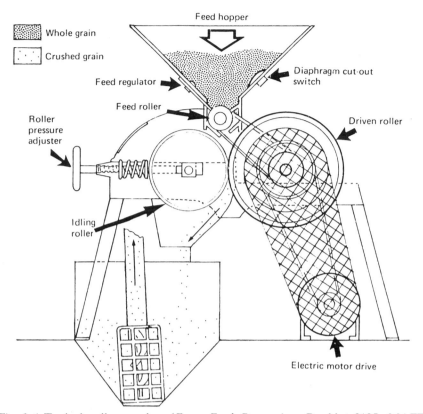

Fig. 1.4 Typical roller crusher (*Farm Feed Processing*, Booklet 2125, MAFF, © Crown Copyright 1989).

Roller crushers

Roller Mills are commonly used for crushing grain, such as barley, for feeding with appropriate supplements to beef cattle. Such mills comprise two smooth steel rollers, held slightly apart by spring pressure. The larger roller is driven by Vee belt from a motor which also commonly drives a small fluted feed roller delivering grain from the hopper via an adjustable feed gate to the crushing rollers. The effectiveness of a roller crusher depends on ensuring an even flow of single grains between the rollers, and this can be difficult to achieve, especially with hard grain where the rollers are of small diameter. It is therefore advisable to choose a machine with larger rollers which more readily accommodate the crushing of larger seeds such as maize or beans. The larger the diameter and width of the roller, the greater the potential output.

The smaller, usually undriven, roller is carried in sliding bearings and can be pressed against the driven roller by a large spiral spring via a hand wheel. The spring acts as a safety device should a hard foreign object be

Table 1.6 Typical outputs in kg/h from roller mills.

Type of crush and moisture content (%)		Mill size (kW)			
		1.5	2.2	3.7	5.5
Light,	14	250	350	500	950
Medium,	14	200	300	450	770
Heavy,	14	150	250	400	600
Light		300	400	550	1100
Medium	18	250	350	500	900
Heavy		200	300	450	750

introduced between the rollers. However, a screen is often fitted in the base of the hopper to collect small objects. The more pressure that is applied to the spring, the more the product is crushed and the lower the specific output. It is important that rollers roll parallel to each other and poor results will be achieved if they are out of alignment. Typical outputs in kg/hr for roller mills are show in Table 1.6.

Just as for hammer mills, roller crushers are readily fitted with automatic cut-out switches of the diaphragm or counterweight type to ensure automatic operation and give opportunity to make use of off-peak electricity. In one version, the crimped roller mill, grooves are cut across the width of the roller, giving a slightly corrugated effect. Here it is usual for both rollers to be driven and then speed can be varied so that one is running faster than the other. Such mills are particularly appropriate for high moisture content grain, such as propionic acid treated barley, as the rollers tend to be self cleaning.

Once grain is rolled it is easily broken up if conveyed by auger to the mixer. Thus, it is common to mount the mill directly on top of a chain and slat inclined mixer, with an intermittent control switch to move the pile of rolled material at intervals from under the mill. Where mechanical conveying is necessary, it is best to use a bucket elevator or slatted conveyor for rolled material or, if need be, a low speed auger. Barley with a moisture content of 18% will generally give a good sample, but where the moisture content is below 16% barley tends to shatter, thus, with lower moisture contents it is normally necessary to moisten the grain. This is typically achieved by trickling water into the auger intake or discharge as the grain is being transferred to the holding hopper. Only 3–4% can be added at one time for barley. It will take some 12–24 hours for water to be absorbed before it can be rolled, thus grain will need to be allowed to soak overnight. Unfortunate consequences of adding water are rusting of the hopper and reduction of storage life. Water should not be added

Plate 1.5 Rotary bale shredder/'tub grinder' (*Cormall*).

to grain held in a larger storage bin as subsequent swelling can cause distortion or even bursting of the bin.

There are more convenient alternatives worthy of consideration. One method is to generate and pass steam through the column of barley raising both its temperature and moisture content, softening it, immediately before it is passed through the roller mill. This saves having to remember to prepare a batch of moistened grain the day before it is rolled and although steam is insufficient to cook the grain there may be, in some cases, a useful partial sterilising effect. But all these practices are unnecessary, anyway, where grain has been stored in hermetically sealed bins or preserved with propionic acid, and with the latter type of treated grain there is the benefit of enhanced storage life of the rolled product.

'Tub grinders'

Increased interest in making wider use of straw has arisen from complete diet feeding of cattle and increased use of mixer wagons. Some hammer mills, such as those fitted to mobile feed mixers, are capable of milling straw, but the main problem is that of shredding and providing a uniform feed to the mill to prevent stalling. Tub grinders, or rotary bale shredders, comprise a rotating tub, the floor of which is fitted with a row of replaceable saw-toothed cutters which shred straw and feed it to a hammer mill beneath. The unit is driven by electric motor or, more commonly, by the tractor pto and mounted on the three-point linkage. In one version, an electronic speed sensor on the main input shaft detects when overload occurs and the rotating hopper is automatically declutched until the load returns to normal. In another version, the hopper is driven hydraulically and its speed controlled by a governor. Such tub grinders are readily suited for big round bales which can easily and quickly be loaded into it by tractor loader.

Mixing

Any feed mixer should thoroughly mix and disperse all ingredients, including minerals and vitamins, to give a mixing effect, not less than the equivalent of hand shovelling, from pile to pile, three times. This should be achieved within 20 minutes. Depending upon design, a certain amount of separation out of some coarser ingredients can occur when mixing is extended beyond the recommended time. Ideally any constituent of 1% or less of the ration should be pre-mixed by hand with a portion of cereal meal before being added to the mixer.

The mixer should be capable of being quickly filled, preferably from top and bottom, and emptied. The thoroughness of emptying is a particularly important design factor when changing from one ration to another. It should be of sufficient volume to accommodate desired batch sizes, usually of one or two tonnes. Since densities of ration vary, the gross volume of the mixer being considered should be checked with the actual volume required. Gross capacity needs to be 10% more than the space occupied by the quantities to be mixed. Pig and poultry rations have a density around 450 kg/m^3, so a 1 tonne mixer needs a capacity of 2.5 m^3, allowing for spare space, and 1.5 tonne and 2.00 tonne mixers need 3.7 m^3 and 5.0 m^3 capacities, respectively.

Mixer types fall into three categories: batch mixers, continuous mixers and liquid feed mixers which may be of the batch or continuous type. Liquid feed mixers are discussed in a later chapter on pig feeding. The new generation of load cell and computer controlled batch wet feed mixers entails that often a dry mixer can be dispensed with altogether, providing a useful saving in capital requirements and running costs.

Batch mixers

There are four main types.

Plate 1.6 Vertical mixer (*Silo & Storage Systems Ltd*).

HORIZONTAL MIXER

This type is no longer commonly used on the farm. It consists of a U-section mixing trough, running horizontally, down which is a mixer agitator comprising paddles or slats fitted to a slowly rotating shaft. This type of mixer can deal adequately with sticky ingredients such as molasses. A thorough mixing is achieved, but at the expense of a high power requirement.

VERTICAL MIXER

This is the most common mixer used in combination with a hammer mill. The mixer is essentially a cylindrical hopper tapering to a cone shape at the base. A vertical central auger driven from the top or bottom at 250–400 rpm, depending on the model, continuously lifts material from the bottom of the mixer to the top. Most makes have a sheath over part of the auger and this is desirable as such a mixer gives a better dispersion and quicker mix (about 8–10 minutes). Without a sheath, mixing takes 12–15 minutes. It is normally recommended that the mixer is started before being filled and run for 10–15 minutes afterwards. Excessive mixing time may lead to a poor mix quality.

Vertical mixers are usually fitted with a bottom intake hopper which feeds material to the bottom of the auger. With a fully free standing mixer this hopper is fitted some 450 mm above floor level, but commonly the hopper is set in the ground which gives a more convenient arrangement for the emptying of sacks into the mixer. The design of installation of below ground hoppers should be such as to facilitate access for maintenance/replacement of the bottom bearing. In addition, the mixer hopper is commonly fitted with a tangential inlet to take pneumatically conveyed grist from the hammer mill. Filter socks are also provided if the air is not circulated back to the mill. When filled in this way it is necessary for the mixer to be capable of starting under full load, and the drive control system should be designed with this need in view. Mixed material is unloaded at the side to a conveyor or bagged off.

Vertical mixers have the benefit of occupying the minimum amount of floor space, giving a vigorous action, and are well suited for meal based rations. They are not suited for liquid addition (such as water and molasses), nor for rolled or flaked material, and are not reputed to be the ideal choice for critical diets (such as pig starter rations) because of some doubts about the accuracy of ingredient dispersion.

INCLINED AUGER MIXERS

These are of two types. In one, a single auger is mounted in the base of a steel bin which is set at an inclined angle and tapers at the base. The auger is driven by a motor at the top via a roller chain, giving an auger speed of 75 rpm. Filling takes place at an intake hopper at the lower end or from the top by gravity, or by pneumatic conveyor. Emptying is achieved by a spout with throttle at the upper end of the mixer.

An alternative machine uses an open, non-tapering body, again set at an incline of about 30°, into the floor of which are set one or two more augers. Such mixers are designed for complete diet and straw mixing with a facility to combine mixed concentrate and forage feed.

CHAIN AND SLAT MIXERS

In recent years sales of this type of mixer have exceeded those of the vertical type by about five times simply because rolling and crushing, or bruising, is by far the most popular method of home milling grain on farms in the UK, and chain and slat mixers give the most gentle action. This type of mixer gives the shortest mixing time (5–15 minutes depending upon size) and it is claimed to be up to 99.9% self emptying, with ready capability to accommodate liquid addition (such as molasses and water). But

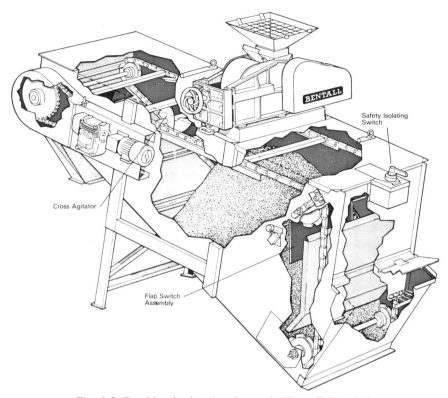

Cross Agitator

Safety Isolating
Switch

Flap Switch
Assembly

Fig. 1.5 Combined mixer/crusher unit (*Bentall-Simplex*).

it has around three times the power requirement, occupies more floor
space and capital cost is more expensive than a vertical mixer. It does,
however, require only around 1.5 m headroom compared to at least 3 m
for a vertical mixer. Accuracy is claimed to be particularly good, with a
dispersion ratio of 1:10 000.

The mixer consists of an endless chain slatted conveyor lifting the
material up on an inclined base towards a segmented agitator, the latter
giving a lateral movement to the ingredients. In one version the chain and
slat conveyor moves along a floor which is flat apart from an inclined
section at one end. Rolled (or hammer milled) material is normally added
by gravity, often with the mill(s) mounted directly on top of the mixer
body. With the latter arrangement, fitting of a flap switch assembly enables
the mixer to be run for long enough to level out the crushed grain, facili-
tating the maximum filling of the mixer and minimising flake damage. A
bottom filling hopper is also provided. Discharge takes place at the top of
the inclined section, often with two parallel outlets, one for sacking and
the other direct to a mechanical conveyor.

Continuous mixing

The general principle of a proportioner, or blender, is that ingredients are simultaneously metered through a calibrated mechanism in pre-set proportions directly to the mill. In this way the mill actually mills and mixes ingredients in one operation before the material is conveyed to buffer storage. The main advantage of such a proportioning system, by comparison to batch systems, is the reduced labour requirement and streamlining of management. However, this relatively loose form of mixing has limitations of flexibility. Thus, the limited number of channels, usually six, does restrict the variety of ingredients that can be included in the ration, unless pre-mix compounds are used.

Accuracy of the mix is dependent upon regular individual calibration because of wide variation in bulk densities of material which can be as much as 10%. Calibration is usually best carried out once a week or following delivery of new consignments. This can be simply undertaken by adapting a plastic bucket. The bucket is filled with 10 litres of water, the water level is marked and the bucket is sawn off at this mark. By filling the bucket with feed, levelling off and weighing, the bulk density can be readily established by simple formula. Thus, a net weight of 4.65 kg gives a density of 465 g/l.

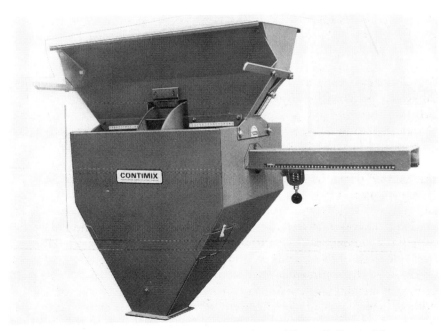

Plate 1.7 Water-wheel type proportioner (*Cormall Contimix*).

Typical mixing accuracies on a well maintained and calibrated plant can range from a claimed ±1% to ±7%, unless special equipment is installed. The addition of mineral and vitamin supplements through the mill is difficult. Not only can mill wear be accelerated by grinding minerals, but grinder heat can partially destroy vitamins, particularly vitamin E. However, in practice the addition of more vitamins can often cure the problem.

There is quite a range of different proportioning equipment on the market. A Danish proportioning system operates on the principle of a water wheel, in that the major proportion of the cereal to be milled and mixed is fed over the main wheel and creates the driving power. Two proportional wheels are attached to either side of the main wheel and the required proportion of the ready mixed concentrate and/or second cereal is controlled by adjustable sleeves.

In another type of proportioner electrically actuated vibratory feeders vary the throughput by one or all of the following means: adjustment to the amplitude of vibration, varying the angle of feed through an incline or adjustment of feed gate height.

The chain and flight proportioner is particularly appropriate for remote fitting at the discharge from bulk hoppers delivering to a common chain and flight conveyor. Each proportioner consists of a calibrated feed slide which controls the amount of material passing from the bulk hopper flowing onto the deadplate of a short section of chain and flight conveyor. Material is moved and allowed to fall onto the common chain and flight conveyor beneath which carries it to the mill. Drive to the proportioner is taken from a star-shaped pick-up wheel powered off the main chain and flight conveyor, thence to the proportioner conveyor by means of a variable speed sprocket and chain drive. Output is controlled principally by changing the drive chain sprockets and fine adjustment is achieved by the slide control.

The most popular type of proportioner mechanism, however, is the variable speed auger. In one version, short parallel feed augers are driven at infinitely variable speeds according to ration. Proportioner units can operate remotely, delivering to the mill intake suction pipe.

One of the best known proportioners on the UK market comprises a unit of up to six channels using short augers with ratchet and pawl drive. Ingredients are delivered from each channel to a hammer mill beneath the unit. Each channel adjustment dial can be set from 0–25. This has the effect of adjusting the throw of the pawl on each stroke: the longer the throw the more ingredients delivered for each movement. A calibration chute can be quickly fitted in place of the back cover to facilitate a collection of samples. Claimed accuracy is 1–2%. An automatic control stops the mill if any ingredient runs out. To avoid deterioration by heating, one channel can be organised to bypass the mill chamber.

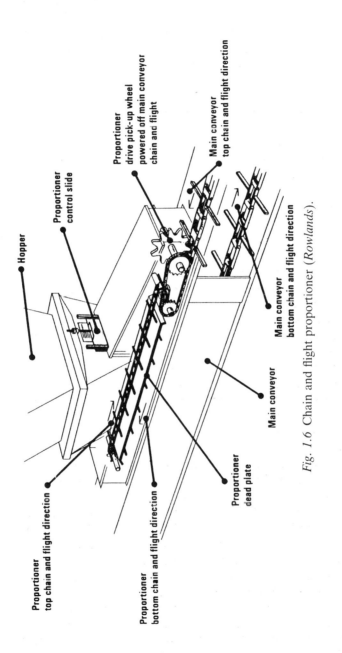

Fig. 1.6 Chain and flight proportioner (*Rowlands*).

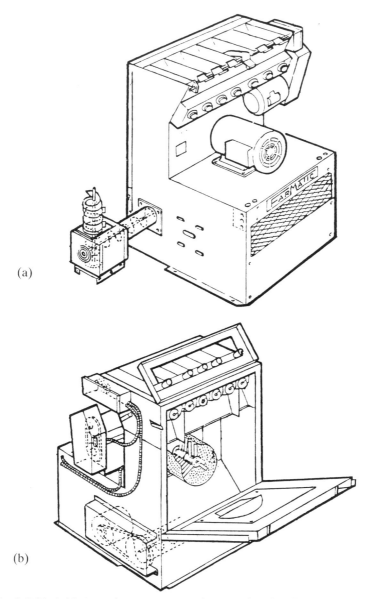

(a)

(b)

Fig. 1.7 Variable speed auger proportioner unit, showing channel adjustment dials and integral hammer mill (*Philco Dierings*). (a) Front view with auger corner. (b) Back view showing inside.

Good flow characteristics are essential for all materials being measured by a proportioning system. It is also vital to achieve consistent measures of ingredients delivered when the settings are varied from one ration to another and back again.

One proportioner recently developed combines the advantages of accuracy of weighers with the speed and convenience of the continuous system. Each unit comprises a hopper with proportioner auger delivering on to a belt weigher that provides on-going verification that the proportioned quantities correspond with demands. Adjustment of the speed of the proportioner augers takes place automatically. A computer controls the whole system which is capable of operating up to twelve ingredients and controlling up to twelve feed rations.

Cubing

There are obvious advantages in pressing the freshly mixed ration into pellets or cubes which are more convenient to handle, with no separation of constituents giving better flow characteristics. It is also likely that there will be less waste and dust, most certainly, will be significantly reduced. Further, palatability is enhanced together with rate of eating capacity. Thus, for dairy cows there are often severe limitations on concentration consumption time available in the milking parlour: when feeding barley and groundnut meal it takes around 3.6 minutes for the cow to consume 1 kg dry matter, but 2.5 minutes to consume 4.8 mm pencils. Cubing or pelleting is essential for some classes of stock, broilers for example, and pigs can convert dry feed more efficiently as pellets than as meal.

However, there is a not inconsiderable price to pay for this added convenience! As well as the introduction of a greater management demanding element into on-the-farm feed processing, a pelleting plant including cuber, buffer bins, cooling and conveyors could well double the capital cost requirement of a basic mill and mix unit and running costs are notoriously expensive, too. The smallest farm cuber costs around £4000 with die rings having a life of 600–1000 hours costing from £300–£700 each, dependent upon cube or pellet size. As output is relatively low (see

Table 1.7 Typical output in kg/h of a farm pelleter (approximate values).

Die ring size (mm)	Cuber size (kW)		
	5.6	7.5	18.6
2.4/3.1/3.5	200	250	550
4.8	250	300	750
8.0	350	400	750
11.0	350	450	1000
16.0	250	400	850

Table 1.7) it is desirable to opt for a larger machine which, for instance, means an investment of at least £9000 for a 18.5 kW cuber. Cube durability also improves with increasing power even within the normal range of farm cubers of 7.5–18.5 kW, and a larger cuber may also have an output better matched to the rest of the plant. Electricity consumption is likely to be in the range 15 kW hour/tonne for 11 mm cubes to 50 kW hour/tonne for 2.4 mm pellets, depending on the material and scale of operation.

The economics are probably just about in balance, but it is the larger scale farm operation which can make a greater success of cubing, maybe even utilising secondhand industrial plant incorporating a steaming process to enhance binding of material.

In principle, the farm cuber or pelleter consists of a heavy cast iron rotating die ring which has tapered radial holes through the periphery. This die ring rotates at around 160 rpm making friction contact with two rollers which, when feed is delivered to the middle, force material through the holes in the die ring. The length of the cube or pellet is dependent upon an adjustable shear bar at the periphery, whereas the cube or pellet diameter is adjusted by changing the die ring. As distinct from industrial plant, which uses steam at approximately 50 kg/hour per tonne of output, farm cubers and pelleters use a cold process, normally requiring the addition of water as a binding agent. If addition of moisture does not prove satisfactory molasses can be added up to around 5% of the ration. Use of

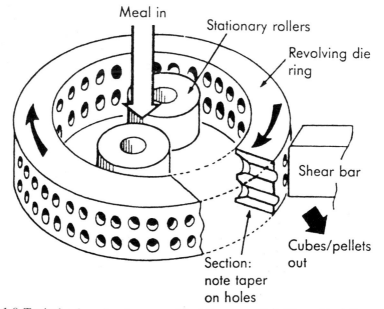

Fig. 1.8 Typical cuber die ring assembly (diagrammatic) (*Farm Feed Processing*, Booklet 2125, MAFF, © Crown Copyright 1989).

steam for larger plant not only increases output but increases die ring life and leads to a higher quality product.

Rations containing a high proportion of fibre – such as oats or wheat offals – do not bind so well and as such should be avoided if cubing is to be carried out.

Once formed, farm produced cubes or pellets are vulnerable in two physical respects: they may be broken up in conveyors and may grow mould and start to crumble. Thus, gentle handling is necessary, avoiding the use of augers if possible. As heat is generated during the extrusion process this must be removed if rapid deterioration of the pellets is not to occur. Where batches of a tonne or less are involved and cubes are to be fed within a few days, all that is necessary is a cooling bin with perforated floor or large inverted V-shaped duct in the base. For larger quantities, forced ventilation is necessary to provide around 0.5 m^3/second of air/ tonne. This can be achieved by bolting on one or more centrifugal fans near the base of the cooling bin. For cubers larger than 18.5 kW, or when live steam is used, specialised high output cube and pellet coolers are almost certainly required.

Control of mill and mix operation

Except on the most basic of mill and mix units it is usual to make use of automatic or semi-automatic controls, which can range from a simple device to shut down a mill when grain supply is depleted to full computer control allowing largely unsupervised operation of the mill and mix plant, thus significantly reducing labour costs and simplifying management. The wider use of computer control generally in industry has prompted the development of relatively low cost sophisticated computer controlled farm mill and mix packages. Further, there are a growing number of companies now offering DIY kits to enable the farmer to convert equipment to various levels of computer control.

Control devices can, for convenience, be divided into three groups of equipment for sensing, actuating and control.

Sensing equipment

For a mill and mix plant this might have the function of detecting:

- when a hopper is empty or full to activate the starting or stopping of a motor or operate an alarm.
- an overflow situation and giving an audible warning.
- a blockage situation and giving an audible warning.
- motion of material to provide an 'all is well' detection system.

Sensing devices commonly used include:

Microswitches – these may be small spring-loaded switches operated by the effect of a light load depressing a lever and can have a current rating up to 10 amps. Alternatively, tiny encapsulated reed switches may be used which are operated by moving a magnet within close proximity and which can carry up to 2 amps only. They might be used typically to count the number of operations of a tipping weigher.

Proximity switches – these do not require any physical contact nor influence of a magnet as they operate on the principle of conductance or inductance.

Pressure diaphragm switches – these are very commonly employed to detect the level of grain in the intake hopper of hammer or roller mills. A rubber diaphragm covers a hole in the lower section of the wall of the intake hopper. Relatively light pressure on the diaphragm depresses a spring-loaded switch behind it.

Flap switches – these are available in a wide variety of types for various applications, typically for detecting flow from conveyors, or level of feed in bins.

Load cells – these are discussed in more detail in Chapter 2. They are of two types: hydraulic or electronic. In the hydraulic type the applied load, for example grain in an intake hopper, pressurises a sealed hydraulic system which links to a weigh scale fitted with pointers coupled to contactors to stop and start augers feeding the hopper according to pointer settings. In recent years, however, electronic load cells are being more extensively used and have revolutionised the control of mill and mix plant. Tiny strain gauges, bonded to steel mountings which are subject to the applied load, detect the amount of stretching or contraction within each mounting and hence the extent of the load. When the tiny wires of the strain gauge are stretched less current can flow and this change is detected by the electronic control system.

Actuating equipment

Sensing devices, like micro and reed switches are not capable of carrying sufficient current to enable direct switching for the size of electric motor normally used for mill and mix plant. Thus, the relatively small current in the sensing device circuit is fed to a solenoid coil to operate the contactors in the motor starter box. When the solenoid coil is not energised the contactors are spring loaded in the open position. The motor starter may also incorporate means of maximising current consumption during start-up conditions.

Control equipment

In simple terms there are three levels of control: manual switching, electrically automated control based on electrical timers and computer control.

For small basic mill and mix plant, manual switching may fit the bill quite adequately, but even simple shut-off switches do provide a considerable improvement in convenience. But one step further, automated switching, using a time clock, does enhance flexibility and reduce labour supervision considerably. In appropriate circumstances timeclock control, too, entails that the plant can be run at night time using cheap rate electricity tariffs with significant savings in energy costs. Sophisticated controllers, using electrically driven timer mechanisms to control the switching of each component to the system, can bring virtual full automation with all the attendant advantages.

However, just as in recent years there have been dramatic changes in the control of domestic appliances such as the washing machine, so too the world of the micro chip is beginning to sweep out the 'clumsy' electro mechanised controllers, giving far greater reliability and enhanced sophistication at relatively attractive cost.

Computer control of mill and mix plant is currently deemed to be cost effective where the throughput is 10 or more tonnes/week, and it is reckoned that as electronic equipment becomes progressively cheaper in real terms the break-even point will reduce. The benefit of computer control could bring forward the time for upgrading of plant on some farms, and it could provide those producers without mill and mix plant with further enducement to do so. But what are the benefits? These can be summarised as: potential savings in running costs, greater accuracy of control and greater sophistication and facility of management control.

In an Electricity Council case history a mill and mix unit providing feed for 250 sows plus progeny to bacon was considered. With a weekly throughput of 33 tonnes and 1675 tonnes annually, up to 75% of the current labour input of 20 minutes/tonne can be saved when the plant is fully automated. Though this amounts to a saving of more than £1/tonne the cost is rarely saved but could be transferred, say, into improved pig management (or more available leisure time!). As far as energy costs are concerned, typically each tonne produced consumes around 21 units of electricity. On a normal day tariff electricity would cost around £1.30/tonne or £2177 per annum, but with opportunity for complete operation running at night time on cheap rate electricity costs would be reduced to around 43.5p/tonne or £728/annum, or a total saving of around £1450/annum.

The tighter control that automation can bring into controlling ingredients by auto weighing can manifest itself in yet further savings compared to many manually controlled systems. Producers are more aware than anyone that an additional 1% protein above requirement could add £2/tonne to ration costs with little or no discernible improvement in feed conversion.

However, it is the added sophistication to enhance management control that sets computer control above electro mechanical automatic control systems. Not only can augers, elevators, mills and mixers be integrally controlled, but the computer controller can be pre-programmed whereby up to maybe fifteen rations, each of say up to fifteen ingredients, can be set into the memory, enabling the operator merely to select the ration number required and press the start button. Furthermore, sophisticated monitoring can be built into the control enabling the plant to be closed down quickly should a fault occur, setting off an alarm and also indicating on a visual display the nature of the fault. As a solid state controller and with less components likely to fail the controller itself is more reliable than the older electro mechanical type.

It is the management information facility that is a particularly useful feature, giving, for instance, re-order indication. This is achieved by the computer counting back from the original start quantity of raw material and indicating on the visual display when a predetermined level is recorded. Some will argue that it is the opportunity for entrepreneurial buying of raw materials for farm feed processing that really gives the biggest savings. By computer it is a relatively simple operation to compare different rations on the basis of price, feed value and protein content.

Choice of mill/mix system

The main factors influencing choice of appropriate system for a livestock enterprise include weekly and annual throughputs, type of stock, the number of different rations likely to be needed and ingredients used, together with labour and finance available. Equipment discussed earlier can be combined in innumerable ways according to one's own circumstances,

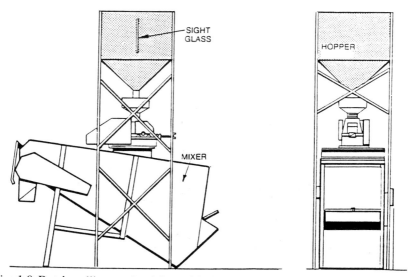

Fig. 1.9 Batch rolling and mixing with feed hopper and volumetric control of ingredients.

but a number of manufacturers do offer complete package systems which simplify design, ordering and installation and should present less hassle when equipment fails.

Some examples of complete mill and mix installations are now discussed.

(1) Batch rolling and mixing with ingredients measured volumetrically or gravimetrically

This is a particularly common combination suitable for relatively low throughputs of a limited number of rations for cattle. As a simple system based on a roller mill mounted directly on top of a chain and slat mixer, it has the benefit of exploiting gravity to transfer milled material to the mixer, low capital cost and such a combination can often be fitted relatively easily into existing farm buildings. Labour needs are in the region of 0.5–0.75 person hours/tonne where the system is manually operated. Up to around 15 tonnes/week can be processed depending on mill and mixer size.

Ingredients are typically controlled by a combination of adding increments of bagged protein supplements of known weight via the lower intake hopper and measuring the cereal component either volumetrically or by weigh hopper. A calibrated hopper with a clear plastic window mounted above the mill provides a relatively cheap means of measuring out batches of grain to be processed. However, variations in bulk density and moisture content and not least the need for the intake conveyor to be

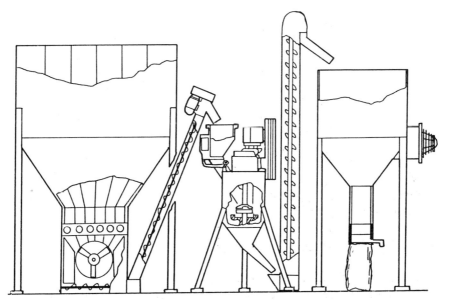

Fig. 1.10 Volumetric proportional system with cubing facilities (*Philco Dierings*).

shut off manually at the appropriate time can all detract from accuracy! A weighing device is a well recommended alternative. This can take the form of either a tipper weigher or weigh hopper, as discussed in Chapter 2. Loading augers or other filling conveyors may be manually controlled or automatically switched off by the addition of a microswitch when the desired weights are achieved.

(2) Continuous volumetric proportioning system

Such a system is appropriate for throughputs of 20 tonnes or more per week for processing of a limited number of rations, with up to six ingredients. Though capital requirement is relatively high, labour needs may be as little as 10–15 minutes/tonne as the plant can be operated unattended for long periods. Volumetric proportioning entails that, with no separate mixer required, it may be opportune to incorporate cubing facilities providing a relatively compact feed processing plant. To minimise risk of the production of unbalanced rations it is essential that the plant is automatically stopped if any ingredient runs out or if the feed supply becomes blocked.

With up to six ingredients required to be bulk supplied to the proportioner care should be taken in design to ensure that the siting of hoppers is well thought out, that as far as possible gravity is used for conveying, and that steps are taken to minimise 'bridging' of materials in hoppers.

(3) Computer controlled fully automated batch system

This is a complete packaged feed preparation system providing from half to two tonnes of feed per hour completely automated for overnight operation. The system comprises a hammer mill with weigher mounted above the intake hopper, and vertical mixer filled from both the hammer mill and from an ingredient dispensing system consisting of auger feed from a holding bin with a further weigher mounted above.

The programmable controller allows for six main ingredients, plus six pre-weighed supplements to be incorporated into any one ration held in the memory. Cereals are delivered to the mill by auger via an oscillating or flip flap weigher, passed through the mill and then pneumatically conveyed to the mixer. Other bulk materials are then weighed sequentially and conveyed to the mixer while microingredients, such as minerals and vitamins, are premixed by discharge into the conveyor feeding the mixer. Proximity switches and the computer system control ensure security of the system, stopping operation and informing the operator should situations occur involving too many ingredients, too many batches, oversized batches or if hoppers are empty. The system will also shut down or sound an alarm for system blockages, incomplete weighing, mill or motor fault and residual material in the hoppers.

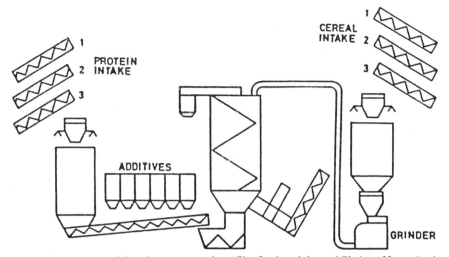

Fig. 1.11 Automated batch system using 'flip flap' weighers (*Christy Hunt Agricultural Ltd*).

(4) Computer controlled fully automated batch system incorporating management information analysis

This system is adopted quite commonly on a number of Danish farms. Raw material is fed pneumatically to the mill via an automatic suction valve which sucks the raw material alternately from four different bins. Material is then blown through a cyclone to a horizontal mixer.

Other materials are added by auger to the mixer which is mounted on three load cells. This provides for accurate weighing of each ingredient in turn prior to mixing and passing to finished meal bins. The visual display provides the operator with clear indications of reasons for stoppages. The programmable controller memory can store four ration recipes and control seven different raw materials. The system also incorporates provision for analysis of management information, such as comparisons of rations, prices, feed value and protein content, as well as re-order indications on raw materials.

References

Electricity Council, Farm Electric Centre (1987) *Feed Preparation on Farms: Milling and Mixing.* Technical Information Agri. 1.

Fisken, J.McN. (1984) 'The economics of home feed processing.' *Agricultural Engineer* **39**(4), 149–152.

Howard, P. (1986) 'On-farm milling and mixing.' *Power Farming* **65**(11).

Leitch, J. (1986) 'Milling and mixing – state of the art.' *Dairy Farmer* **33**(8), 17–21.

Lewis, R.N. (1987) 'Computer controlled on farm feed preparation. "Pork and

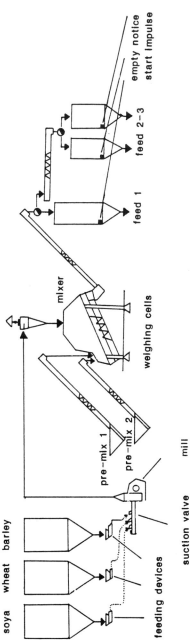

Fig. 1.12 Automated batch plant using load cell weighing of materials into the mixer (Kongskilde Skiold).

Chips"' Conference paper. Suffolk Coastal Pig Discussion Group and New Buckenham Pig Club.

MAFF (1989) *Farm Feed Processing*. Booklet 2125.

Mitchell, D. and Kneeshaw, A. (1987) 'Home mixing for livestock.' *RASE Reference Book and Buyer's Guide*, 217–219.

Pierson, S. (1987) 'On farm feed preparation.' *Farm Buildings Association Winter Conference Report*, 9–13.

Summerfield, A.J. (1984) 'Feed preparation on the farm.' *Agricultural Engineer* **39**(4), 139–140.

2 Feed storage, handling and measurement

Storage – general considerations

Type of material

The density of the various forms of concentrate feed used has a significant effect upon storage volume requirements, with, for instance, unmilled beans and wheat needing less storage volume than barley or oats. Tables 2.1 and 2.2 provide a guide to the approximate storage volumes required by various feedstuffs. The actual volume taken up will vary considerably from sample to sample of the same material, depending on 'bushel weight'. Note, too, that milled materials occupy 20–25% more space.

Besides density, flow characteristics need to be taken into account when determining the size and type of storage bins needed, as uninterrupted flow is a vital requirement of any materials handling situation. Bridging of feed in bins and chutes is a common problem, the risk of which, with some precautions, can be largely overcome. Good flow performance will be

Table 2.1 Approximate space in m^3/tonne occupied by cereals and pulses in bulk.

Crop	Unmilled	Meal*
Wheat	1.3	1.9
Barley	1.4	2.0
Oats	1.9	2.4
Beans	1.2	1.8
Peas	1.3	1.7
Maize	1.4	1.7

* Rolled, as opposed to ground, material normally occupies more space than these figures indicate

Table 2.2 Approximate space in m³/tonne occupied by various other feedingstuffs.

	Feedingstuff	Space occupied
'Straights'	Fishmeal	1.7
	Soya bean meal	2.4
	Groundnut cake	1.8
	Dried grass meal	4.0
	Flaked maize	4.0
	Dried beet pulp	4.0
	Weatings	2.7
Protein concentrates	Meal form	1.7–1.8
	Pelleted	1.4–1.5
Finished rations	Meal/mash	1.7–2.2
	Pellets/cubes	1.4–1.5
	Mixture containing large percentage of rolled materials	1.9–2.8

aided by a low angle of repose, that is the natural angle to the horizontal made when material is formed into an unsupported heap.

Dry, clean unmilled cereals and pulses normally flow quite easily and hoppers with bases sloping at not less than 45° to a central or side outlet should empty normally without trouble. However, damp grain and impurities increase the angle of repose and thus for wheat at a moisture content (m/c) of 10% the angle is 27° but at 20% m/c it is 37°. Similarly, milled cereals, mixed finished rations and premixed concentrates flow less easily and steeper slopes are required for unassisted complete emptying of bins. Hoppers with one vertical side and the sloping side angled at not less than 60° to the horizontal should ensure successful emptying. Should bridging still present a problem an electrically operated vibrator plate can be incorporated on to the side of the bin just above the outlet. This can operate automatically, if need be, such that when interruption of flow is detected the vibrator starts up and shakes the material down, thus restarting the flow.

The greatest potential bridging problems occur with rolled and flaked materials and also wheat offal and fishmeal. Here, a number of possible remedies can be adopted, such as steeper sloping sides coated in, or made of, a low friction material. Hence bins may be constructed from fibreglass or a PTFE plastic lining may be employed. Special auger discharge bins are also available, consisting of two sloping sides down which material slides to a horizontal trough which carries an open flighted large diameter slow speed auger.

Holding capacity

·How much storage capacity to provide for concentrate feed or raw materials is not an easy decision. Capital investment on bin storage can be substantial with typical bin storage costing £60–100 per m^3, to which must be added the financial outlay to purchase the feed being stored. Thus, the adoption of the current industrial manufacturing practice of 'just in time' supply of material can reduce overheads but it is vulnerable to supply delays! Annual and peak feed requirement will dictate the minimum amount of storage necessary which, for mixed rations, should never be less than 2 to 3 days supply for home mixed feed and certainly much more for bought-in concentrates. The shelf life of some materials, such as fishmeal, will limit how much it is desirable to store at any one time.

Nevertheless, it is generally unwise to skimp on storage capacity. Not only can flexibility be enhanced and management hassle be reduced by the provision of sufficient storage space, but sometimes opportunities are created for entrepreneurial purchasing of materials. Recent instances have been quoted where such practice has made all the difference between profit and loss in the somewhat volatile pig markets of recent years.

Storage of raw materials for on-the-farm processing

For the majority of farm feed processing units most cereals will be home grown and handled and stored in bulk. Here there are advantages in locating the mill and mix plant adjacent to the grain storage facilities where cereals can often be conveyed direct to holding bin(s) in close vicinity to the milling plant. Such bins may need to hold as little as one day's supply of grain.

Longer term storage of bulk grain/pulses

Bulk longer term storage of grain is generally organised in one of three ways: an on-the-floor system, bin storage within a grain store or in the form of free standing bins with weathertight roofs. In general terms, the latter type of bin reduces overall investment costs by avoiding the need for building accommodation. But a nested square bin system under one roof, though requiring greater capital investment, provides the greatest flexibility and lowest labour needs.

On-the-floor storage, however, is normally the cheapest form of storing grain, but by comparison to a bin storage system unloading is often less convenient, and especially so when it comes to ensuring regular flow to the feed processing plant.

Long term bulk grain bins are available either as independent free standing round bins or nested in groups using square or rectangular bins. Circular bins provide economy in having the smallest wall-to-content ratio of any shape. The use of corrugated or pre-formed ribbed sections of suitable diameter produces a robust structure yet utilises only light gauge panels, saving on material cost and simplifying construction. Sizes vary typically from 3.0–8.0 m in diameter, 2.5–7.25 m in height and 13–200 tonnes capacity, or more. Often such bins are fitted with low volume ventilation equipment or with ducting and fans, which make drying possible.

Bins are filled through a hatch at the apex of the roof. A separate access hatch is normally also provided in the roof assembly. A roof ladder with hand rails and platform is an integral part of the bin.

Where bins are arranged in an arc around a central reception pit, a reasonably effective handling system results. Each of the bins empties by auger into a central sump connected to the reception pit. Final emptying can be by radial sweep auger, which comes into operation once the angle of repose of the grain is reached, to sweep the remaining grain into the central sump and thence to the reception pit, or shovelling by hand into the sump. A long auger, mounted on a wheeled chassis, runs from the reception pit and can be used to fill any of the bins or to outload into lorries, trailers or supply the feed processing plant. With the necessity for a regular and frequent supply of grain to the latter it is advisable that the reception pit is protected from the weather.

In common with all circular structures care must be exercised in filling and emptying a round bin storage system to avoid uneven pressures on different sections on the side wall. Thus filling takes place centrally from the roof, ideally with a grain spreader to ensure uniform filling around the circumference. In most cases unloading also takes place from the centre of a normally flat floor.

Low cost circular bin kits are available for DIY installation, often on a short or medium term basis within an existing farm building. These may comprise rings of weldmesh lined with bitumen impregnated paper or a more substantial popular version comprises bins made up of sheets of oil tempered hardboard bolted together into a circle and reinforced by circumferential straps. Access hatches, auger tubes, inlets and ladders are attached to provide an open topped structure typically 2.4–3.7 m high, of diameter 3.4–6 m and with capacities of 17–84 tonnes. Filling and emptying must be carried out centrally, emptying being achieved by a central auger position through an inclined side inlet at 45°. Low volume ventilation equipment is often fitted in order to stabilise grain temperature and thereby extend the safe storage period, while low volume warm air drying units are also available.

Such low cost storage bins are a popular means of storing grain destined for processing through a crusher and horizontal mixer for the preparation of cattle rations.

When nested in groups round bins do not make the most effective use of floor area. This may be of little importance outdoors, but the confines of an existing building may promote the adoption of square bins instead. Square bins are normally based upon simple modular pressed panels of galvanised steel bolted together to form single or banked bins housed within an existing building. Alternatively, the bins themselves can form the walls of a new building with the modular panels extended upwards to act as the support for a roof. By selection of the appropriate number of panels, bins can be constructed from around 2 to 5 m square and 3–6 m high to hold from 13 to over 112 tonnes of grain. Bins are normally arranged in two rows with an overhead catwalk.

Bin floors are normally perforated and served by underground air ducts to provide an in bin drying system and, in some cases, an airsweep floor system is incorporated to remove the last of the grain from the bin when it is being emptied. Filling of bins is undertaken using a common overhead belt or chain and flight conveyor fed from a bucket elevator collecting grain from a grain reception pit with swivelling outlets to direct grain into the appropriate bins. A similar conveyor system running just below and between the row of bins is used for emptying.

On-the-floor storage provides one of the cheapest and most popular forms of storing grain. The requirements are relatively simple: a weather-tight and preferably bird proof building with sufficient height and good access plus, normally, a drying system based on above or below ground ducts and a fan system. Filling is achieved by a materials handler or tractor fore end loader, or by auger, belt conveyor or grain thrower, with grain heaped over the lateral ducts to a height of around 2.4–3.6 m then the top levelled. Emptying is achieved by similar loader bucket, auger or belt conveyor.

Fundamentally, the most important aspect for on-the-floor storage is to ensure that the surfaces supporting the grain are able to withstand the considerable lateral thrust. Thus, a conventional concrete block wall could easily fail without adequate provision being made by reinforcement or, better, the use of free standing grain walling, L-shaped in section, designed and braced to take the thrust of the grain being supported. Such panels can also be used as partitions between consignments of grain. Alternatively, suitable pre-fabricated metal walling can be used where the stresses are transferred to special stanchions rather than to the normal structural stachions or walls of the building.

Alternatives to storing dry grain – which is to be fed to stock

Where grain is to be fed to livestock it may be possible to store safely at higher moisture contents. To a modest extent this can be achieved by reducing grain temperature using a low volume ventilation kit for safe grain storage over a few weeks. Grain at 20% moisture content can be safely stored for stock use without mould formation for 2 months or so, as long as the storage temperature is 7°C or less. However, higher moisture content grain can be stored using sealed silos by treating the grain with propionic acid or caustic soda. The disadvantages are that once grain is stored under these conditions, one is more or less committed to feeding it to stock and it is unlikely that it can be sold for other purposes – certainly it will be no use for seed or milling, as the germination will be impaired.

However, there are obvious advantages for storing higher moisture content grain, as harvesting can be started earlier, the cost of drying can be eliminated, no water will need to be added for rolling and palatability of the ration is enhanced, while only small changes occur in its feeding value. Moist grain is most appropriate for feeding to cattle for when it is finely milled for pigs or poultry the specific output of the hammer mill is reduced. In one trial, the growth rate and FCR of pigs fed on barley stored in a sealed store was impaired by 5.2%. However, in another trial the performance of pigs fed on acid treated grain was the same as for pigs fed on dry grain.

Sealed storage

Grain is contained in a (supposedly) airtight silo in which the depletion of oxygen from its normal level of 21% during the process of grain respiration results in the suppression of organisms which might otherwise cause grain to deteriorate. The system permits the safe storage of grain at moisture contents of 18–24% in tower silos for which the dimensions are typically 4.5–8 m diameter and height up to 20 m.

The achievement of a gas tight store demands a very high standard of silo construction. The most commonly used material is vitreous enamelled steel panels bolted together over a mastic seal and with an aluminium alloy roof structure. Filling is achieved by blower or auger through an airtight filling hatch. Emptying is achieved by a sealed unloading system normally consisting of an under floor discharge auger with rotating sweep auger to remove the last remnants of grain. Once removed, grain should be used in the shortest possible time in order to minimise heating and deterioration.

The success of this technique depends on preventing or minimising air leakage into the sealed chamber. This can occur as a result of changes in internal and external air pressure caused by changes in air temperature. Pressure differences due to diurnal temperature changes can be counter-acted by using a 'breather bag' positioned within the air space above the grain, but ducted to atmosphere, to enable the air pressure to remain constant. Breather bags are preferable to pressure relief valves for press-ure compensation as they can accommodate pressure changes without allowing the entry of air which would enrich the store atmosphere with oxygen. Nevertheless, breather bags cannot accommodate severe changes of temperature and a breather valve should be fitted as well.

Inevitably, air will also leak in when unloading, the effect of which will not normally be great during winter months, but could pose a serious problem in warmer weather. Thus the more frequently grain is removed, the more air will leak in.

As the atmosphere inside the store is extremely low in oxygen the store should never be entered without the wearing of breathing apparatus and full safety precautions being observed. Even when the store has been emptied and ventilated any person entering should wear a safety harness and lifeline with two attendants outside who could pull him or her to safety if problems occur.

Chemical treatment

Propionic acid treatment of grain for storage is a long established technique which works well, the justification of which is often influenced by the cost of acid compared to the cost of drying. Grain is passed through an acid applicator comprising hopper and mixing and discharge auger into which a metered supply of acid is pumped at typically 6–10 litres/tonne grain, depending on grain moisture content. Propionic acid is volatile and it is recommended that to prevent the active material being removed from the grain it should not be conveyed pneumatically for at least 24 hours after treatment. Treated grain is normally stored on the floor or in hardboard bins, and it is not suitable for storing in metal bins. Potential corrosion problems within the feed processing plant need to be assessed before adopting the technique.

An alternative chemical treatment is the use of caustic soda which, when absorbed by the grain, disrupts the seed coat and facilitates the entry of ruminant digestive enzymes and micro-organisms into the grain. The digestibility of whole treated grain has been proven, in feeding trials of stores, to be at least as good as for rolled grain.

Care needs to be taken when handling both chemicals to avoid skin and eye contact and protective clothing should always be worn.

Short term bulk storage of raw materials

Cereals

Where the grain store is located adjacent to the feed processing plant, provided that there is an efficient conveying system cereals can be extracted from long term storages bins directly to an appropriately sized hopper above the mill. But where grain is bought in, or if the grain store is remote from the processing plant, then provision should be made for certainly not less than one week's storage capacity, or sufficient space to receive at least one whole bulk load of grain. Generally, in order to

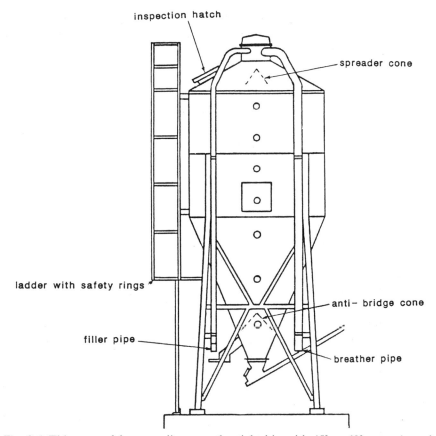

inspection hatch

spreader cone

ladder with safety rings

anti- bridge cone

filler pipe

breather pipe

Fig. 2.1 This type of free standing, weathertight bin with 45° or 60° cone shaped base is recommended for holding crumbs, pellets or cubes (*Collinson*).

achieve a sensible buffer, it would be wise to accommodate around a month's supply or so of grain. Thus, with a mill and mix plant having an annual throughput of approximately 500 tonnes, or about 10 tonnes/week, two 15–20 tonne, cereal bins might be provided. Ideally such bins should be self emptying. The type of bin shown in Fig. 2.1 can serve the purpose well. Here filling can be achieved pneumatically or, where the height does not exceed 8 m, by long auger. Side discharge makes it possible for cereal to be delivered by gravity through an exterior wall to the grain mill. Setting the bin on to an adequately designed concrete plinth can assist discharge reach inside a building. Alternatively the bin legs can be extended, within limits. Cereal bins sited outside the feed processing plant building facilitate access for filling, give less clutter inside and leave more space for the plant itself and other storage.

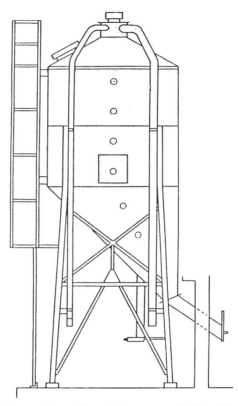

Fig. 2.2 Holding bins such as this which are available with cone shaped base steeply sloping at 67.5°, with central or side discharge, are suitable for meal storage (*Collinson*).

Non-cereral ingredients

With annual throughputs of 500 tonnes or more it is feasible to store non-grain ingredients in bulk hoppers of the type shown in Fig. 2.2 or in special auger discharge bins of the type discussed earlier. Indeed, bulk storage is a prerequisite for any automated feed processing plant. Thus, for a pig unit of 250 sows and progeny to pork/bacon weight consuming approximately 1000–1100 tonnes/annum, up to four non-cereal ingredient bins might be justified, three of which, for example, with a capacity of 25–35 m³ for ingredients such as soya extract and wheatfeed, and perhaps a smaller one of approximately 10 m³ to hold fishmeal.

Nested or compartmentalised bins are available to give typically two or four separate, adjacent, smaller compartments. These are particularly useful as a convenient means for bulk storage of smaller quantities of raw materials or mixed rations. Furthermore, the close proximity of compart-ment outlets facilitates conveying arrangements into the feed processing plant or distribution of mixed feed to stock.

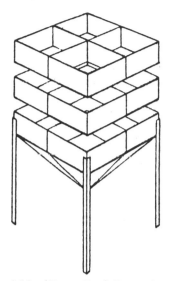

Fig. 2.3 Compartmentalised bin (*Farm Feed Processing*, Booklet 2125, MAFF, © Crown Copyright 1989).

Precautions should be taken to ensure that ingredients held in bins are kept as dry as possible in order to minimise risk of bridging and deterioration.

All bought-in bulk ingredients may be delivered from tipping vehicles or, most commonly, from lorries incorporating pneumatic conveying facilities. Here, all that is necessary is to fit suitably sized fixed inlet and outlet pipes to storage bins. But for larger plant it may be appropriate to install a reception pit and mechanical conveying in order to provide greater flexibility in choice of supply. Adequate access should be allowed for delivery lorries, of course, with sufficient storage capacity to cope with delivery of one complete vehicle load.

Short term storage of raw materials in sacks

The keeping quality of many non-cereal ingredients can limit their period of safe storage from 2 up to maybe 25 weeks depending on material, moisture content and ambient temperature. Where such materials are used in small quantities only, it may well be impracticable and uneconomic to store in bulk and handling and storage in sacks become inevitable.

Logically, sacks should be stored so that they can be conveniently and quickly emptied into the mixer. Floor area required will depend upon stacking height and will range from around 0.9 m²/tonne for dense ingredients such as fishmeal up to 2.0 m²/tonne for bulky materials such as flaked maize or dried sugar beet pulp. Ideally, space and access should be sufficient to allow for sacks to be delivered, unloaded and stored on pallets in order to minimise unnecessary manhandling.

Storage of processed concentrates

Whether concentrate feed is home-mixed or bought-in, bin storage is crucial – enough is needed to provide sufficient buffer capacity to ensure that stocks will not run short, but not so much that feed is likely to deteriorate. As discussed earlier, 2–3 days' capacity may be sufficient for home-mixed concentrates. This should ensure that the plant does not need to be worked at weekends but it does not give much scope to deal with breakdowns. At the very least, 10 days' supply is needed for bought-in concentrates, longer if there is the probability of disrupted supplies. It is desirable that bins should hold about 25% more than a normal lorry load so that deliveries can be conveniently and economically organised. For bought-in concentrates some feed companies offer special low interest finance deals to fund the purchase of storage bins.

Free standing self emptying storage bins are normally sited just outside the livestock house or milking parlour and filled pneumatically from the delivery lorry or by auger. Feed is normally conveyed into the house by chain/cable and disc conveyor or centreless auger. Suitable bins incorporate steeply sloping bases of not less than 55°, as discussed under storage of non-cereal ingredients, an example of which is shown in Fig. 2.2.

To deal with potential bridging problems, when material may become damp, a vibrator plate can be bolted above the hopper exit point and brought into operation manually or automatically when flow ceases temporarily.

Special provision may be necessary to store home-produced cubes or pellets with storage bins fitted with ventilated floors or cooling fans.

Small quantities of feed may not justify bin storage and here rations can be contained and handled in sacks, plastic bins or trolleys.

It is important that the bin incorporates a dispersal cone or plate at the top to prevent blown feed from sticking to any particular section of the bin wall. The pneumatic delivery pipe (where fitted) should not be less than 100 mm in diameter and should have an exhaust pipe of not less than 150 mm diameter to permit excess dust to be collected and to minimise risks from condensation. An inspection hatch, complete with internal and external ladders and safety rings, is normally incorporated to permit feed checks to be made and to allow periodic cleaning. A feed replenishment flap switch is a useful adjunct to give a convenient warning of feed running low.

Handling systems for concentrate feed and raw materials

The handling system most appropriate for any livestock enterprise is influenced by type and quantity of rations, whether feed is home-mixed or bought-in, type of stock, method of feeding, distances for feed to be moved and labour and capital available. In general, the larger the quantities to be handled the greater are the opportunities to reduce labour input and drudgery by handling in bulk. Inevitably, there has been a marked trend towards wider adoption of bulk handling methods, the benefits of which should not least allow the stockperson more profitable time for stockmanship. But even for small enterprises it may often be possible to streamline manual handling systems by applying simple work study principles and questioning whether there are simpler, more effective means of moving feed around the farm.

Whatever handling system is used it is vital that it does not impair feed quality. Thus, good design should ensure that feed is not trapped and allowed to deteriorate nor should it be possible for feed to separate out and become unmixed. Similarly, provision should be made to minimise risks of contamination from vermin and wetting by rain. For rolled and flaked materials, and also for cubes and pellets, there is the danger that certain conveying equipment could cause disintegration – not only impairing the physical condition of the feed but also promoting dust and waste. In particular, conventional augers and pneumatic conveyors are prone to this problem.

In general terms, farm feed handling is organised on the basis of either a mobile system, where feed is moved in discrete packages ranging from those which can be manhandled to consignments of several tonnes, or a fixed bulk feed conveying system using mechanical conveyors or conveying in liquid form. Quantities involved and distance to be moved will largely influence which is the more appropriate system for any enterprise.

Mobile systems of handling

These systems are particularly appropriate where distances involved are so great as to rule out the feasibility of using mechanical conveyors, and where relatively small quantities have to be dispensed to a larger number of scattered destinations, such as to yards of barley fed beef cattle. Labour requirements are inevitably relatively high, depending on size of consignment, type of transport used and distance. Such systems can conveniently be grouped into two types: those based on full or partial manual handling of small units of feed and those using mechanised bulk transport, which can significantly enhance productivity.

Manually based systems

Traditionally it has been the practice to handle concentrate feed in sacks, mostly in the form of paper bags, whether as bought-in or farm-mixed rations. For smaller quantities of particular rations the use of sacks is almost inevitable. Though labour needs are high, sacks do give advantages of easy rationing and opportunity to distribute feed direct to stock in feeding troughs. Alternatively, as paper sacks tend to tear, small quantities of home-mixed rations can be handled in light plastic containers such as plastic dustbins, which in some instances make for more speedy and convenient handling and distribution of feed. Bins can be calibrated and marked up to assist rationing.

Sacks or plastic bins can be transported by trailer, or more commonly nowadays by pallet using a rough terrain fork lift truck, which not only provides more flexibility in distribution but can also minimise manual lifting for loading and sometimes for filling high feeder hoppers.

Wheeled feed barrows and trolleys are a common, useful and low cost means of handling dry feed for manual, rationed, dry feeding systems, for instance, in small and medium sized piggeries. The trolley should hold as much feed as it is practicable to push by hand in order to minimise replenishment time.

Bulk transport systems

Larger unit consignments save time, and the marked trend in recent years to bulk handling round the farm has been facilitated by the ubiquitous use of rough terrain fork lift trucks. Telescopic boom materials handlers, and especially those with four wheel and crab steer, offer a particularly flexible means of collecting, transporting and delivering feed round farm build-

Plate 2.1 A telescopic boom materials handler provides useful reach to fill hoppers and feed mangers (*Sanderson*).

ings. They are extremely adaptable, provide a wide choice of feed handling method such as by pallet forks or by bucket and the capacity of some of these machines is as great as 1.5 tonnes. When one is used as a pallet handler, feed can be contained in sacks, plastic bins or in portable self emptying bins, filled at the feed processing plant before being delivered to the stock unit, and mounted on temporary legs. In some instances such portable feedbins are transported by trailer. Alternatively, feed may be transported in tipper boxes while some farmers use reusable 'big bags' from which feed is off-loaded by release of a cord tie at the bottom.

On a number of intensive livestock farms large quantities of feed have to be transported to widely separated stock units, where feed will either be transferred to bulk bins or fed directly to the cattle. Farm trailers can be adapted for the former purpose, with unloading by auger or by using a tipping trailer with tailgate fitted with pneumatic conveying facilities to blow material into storage bins. In more recent years specialised bulk feed delivery transporters have become available, fitted with a large inclined auger to deliver feed to the required point. On some, the delivery auger can be hydraulically extended making it possible to fill quite high storage hoppers. Such concentrate hoppers are of two main types: tractor mounted or tractor drawn. In addition, complete diet feeders provide ability to handle admixtures of concentrate and forage feed (discussed in

Plate 2.2 Tractor mounted concentrate feeder (*Newlands*).

Plate 2.3 Bulk concentrate transporter with auger delivery for filling high bulk hoppers (*Rowlands*).

Chapter 3). Tractor mounted concentrate transporters have a capacity of around 0.5–1.5 tonnes and are suitable for delivering small quantities to ad lib feed hoppers or direct to feed mangers.

Larger trailed bulk feed transporters typically have capacities of 3–12 tonnes. In some cases the transporter includes two compartments which can save on journey time when two different rations have to be delivered to a similar destination. Alternatively, it is possible to use the transporters for mixing, say, rolled barley with purchased balancer feed by metering out in chosen proportions.

Fixed conveying equipment

Where feed transport distances are relatively short, installation of fixed conveying equipment can save considerably on labour, particularly with automatic operation. Feed can be conveyed in liquid form mixed with water or whey, as used for pipeline feeding of pigs, or more commonly in dry form using mechanised conveyors. With careful planning it may be possible to minimise the number of separate conveyors used, each of which may require its own drive motor, and also to exploit gravity as being the cheapest means of conveying.

There is a wide range of mechanical conveying equipment suitable for handling grain and other raw materials as well as meal, cubes and pellets in, and without, almost any farm building complex. Factors to consider when selecting appropriate conveying equipment include capacity, power requirement, reliability, ready ability to accommodate changes of direction and level, ease of cleaning and self cleaning ability, noise levels, dust generation, risk of damage of conveyed material and, not least, cost.

Augers

Augers remain the most commonly used conveyor as they are simple, versatile and relatively inexpensive for conveying over short distances in any plane from horizontal to the full vertical. In conventional form they are normally limited to horizontal lengths of 30 m per driving head, but this can vary with manufacturer. But they are somewhat noisy in operation and prone to induce damage in susceptible conveyed materials. In conventional format they comprise a tube within which rotates an Archimedian screw or scroll of continuous metal flight welded to a shaft. The shaft is driven through a V belt drive from an electric motor mounted on top of the auger tube at one end. Alternatively they may be driven by a small engine or hydraulic motor coupled to the tractor hydraulic system when used for unloading bulk feed transporters or mobile mill mixer trailers. Small sizes typically come in lengths up to 7.5 m and shorter lengths are easily moved by one person. Longer versions are available mounted on a wheeled chassis with means of adjusting auger angle for filling (or emptying) of high bins. Versions are also available fitted with a sweep collector for picking up grain from floor stores.

The speed of rotation varies in the range 700–1400 rpm and for a 115 mm diameter version 975 rpm would be typical. Auger diameters available include 76, 89, 102, 115, 127, 152 and 229 mm. In general terms, for

Table 2.3   Auger outputs in tonnes/h at various angles for a throughput of clean barley at 17% moisture content.

Diameter (mm)	Angle of inclination (°)						
	0	15	30	45	69	75	90
102	16.0	12.7	11.2	9.9	8.7	6.4	5.2
152	41.1	36.8	32.5	27.9	23.1	17.9	12.4

an increase in diameter of about one-third the output is increased by almost three times. Actual throughput of any particular size will depend on type of material conveyed, its moisture content and the angle of inclination. When used vertically, auger output is about one-third that when used horizontally. Table 2.3 provides an indication of typical auger outputs at various angles.

Feed adjustment is achieved by means of an adjustable slide over the feed intake, but most often this slide is left fully open giving the possibility of motor overload with high moisture contents and high angle of inclination. In order to avoid the possibility of the motor burning out it may be worthwhile to specify for the auger to be supplied with an oversized motor. Some augers can be easily adjusted in length by adding or substracting sections of tube and internal flight, but here it is vital to check that the motor is capable of the extra power needed to convey over extended lengths.

Augers are relatively easy to clean out. If excessive wear is to be avoided they should not be run empty. The size of clearance between auger flights and the casing may vary in the range 6–20 mm and has a particular impact in respect to potential damage to conveyed material. For vulnerable materials it is advised that this clearance should be at least 12 mm. There is little effect on throughput with small clearances, but larger clearances reduce specific power requirement. Conventional augers are not really suitable for the handling of rolled or flaked material which is easily broken up. However, some improvement can be made by altering the pulley and belt drive to reduce auger speed to around 400–500 rpm.

Three special forms of auger are worthy of note for moving grain or concentrate feed. High capacity augers mounted in an open topped U-trough are available which not only permit a number of materials to be fed into the conveyor along its length but also can minimise bruising of material with a high moisture content. They can also be used for conveying of silage and other forage feeds.

Conventional augers are best used for conveying feed in straight lines, any change of direction would have to be accommodated by two or more separate augers. However, a form of auger is available using a bevel gear

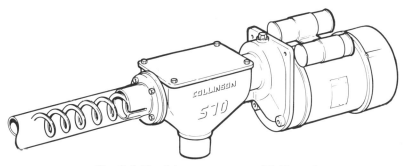

Fig. 2.4 Flexible auger system (*Collinson*).

drive to make up to three right angle turns (Devayor). Alternatively, the centreless auger does provide the flexibility (literally) in accommodating changes of direction. This comprises a continuous spring-like steel spiral, driven one end and mounted in a uPVC flexible tube or galvanised tube with uPVC bend sections. In such a way, with slow curves, a 90° change of direction can be achieved within a 3 m radius. These augers are frequently used for moving meal, pellets and cubes, with minimal damage, from storage bins to livestock feeders over distances up to 120 m with a single motor drive. This type of auger is further referred to in chapters on pig, poultry and cattle feeding.

Chain and flight conveyors

These comprise an endless, sprocket driven chain carrying transverse flights which drag the material along a wooden or metal trough and which may be enclosed or open topped. The latter versions operate between the horizontal and slopes of about 10°. When completely enclosed they can elevate up to an 80° incline and here fabric flights, attached to the conveyor chain, fill the trough cross-section and inhibit material from running back when inclinations are steep.

When 'cleaning flights' are fitted such conveyors are virtually self cleaning and when used horizontally they may be reversed easily, though possibly with reduced output.

Flight widths of up to 300 mm spaced 150–300 mm apart with chain speeds of 0.1–0.77 m/s, are used to give outputs up to 30 tonnes/h. A 1.4 kW motor will power a 30 m length at 30 tonnes/h.

Chain and disc/cable and disc conveyors

Tube conveyors are widely employed in moving concentrate feed to feeders from bulk holding bins in a ring circuit, and are discussed more fully in Chapters 3 and 6 in relation to pig and poultry feeding.

They comprise steel tubing of 35–60 mm diameter through which passes either an endless chain, steel cable, or nylon rope fitted with round nylon discs driven by an aluminium or nylon sprocket wheel in a drive unit, normally with an automatic tensioning device. Special corner units, comprising a wheel or sprocket fitted inside a pressed or cast shell, enable complex circuits to be negotiated horizontally, vertically or at an angle up to a maximum of 500 m in length with some systems. The normal maximum length expected would be 300 m. Typical chain or rope speed is 24–30 m/minute using a 1.5 kW motor. Throughput capacity expected will be 450–2000 kg/h depending on the tube diameter and the speed. Thus, for example, a 38 mm tube conveyor at 35 m/minute gives 400–500 kg/h and a 60 mm tube at similar speed gives about 2000 kg/h.

Because these conveyors operate in a ring circuit they are most appropriate to supplying feeders, or poultry feed troughs, arranged in a double row or on four sides of the floor layout. These conveyors can handle all types of concentrate including pencils up to 10 mm in diameter. Quiet, smooth operation is a particular feature of this type of conveyor.

Belt conveyors

Belt conveyors comprise an endless flat belt up to 300 mm wide often operating in a dished shape at speeds up to 1.27 m/s giving outputs up to 50 tonnes/h over distances up to 40 m. Power requirement is approximately 1.4 kW per 15 m length. They are normally used in the horizontal plane although up to 15° inclination is possible. By fitting a ribbed belt, delivery can be achieved up to a 30° inclination. Belts may be reversed in direction.

Belt conveyors are particularly appropriate for conveying grain but also any material easily broken up, as they cause no damage and also comparatively little dust is raised. They can be used for conveying forage feed too. Complete self emptying is possible and loading and unloading can be achieved anywhere along the length. Feed-on hoppers are provided to prevent material bouncing off the belt and take off points are easily movable. They do tend to be more expensive than chain and flight conveyors.

Pneumatic conveying

Pneumatic conveying provides a highly flexible means of moving material both horizontally and vertically, using a single conveyor only, around the type of tortuous route often associated with old converted farm buildings. But such flexibility is provided at considerable cost – both capital and running costs – especially where outputs of more than 10 tonnes/h are required, but in particular running costs are very high. Moving the conveyed material in considerable quantities of air is an inefficient process

SUCTION-BLOWING SYSTEM

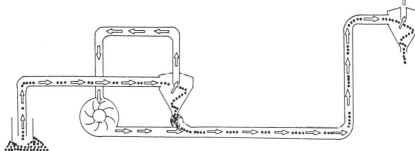

The suction-blowing system is capable of conveying material from several points of intake to differing outlets.

SUCTION SYSTEM

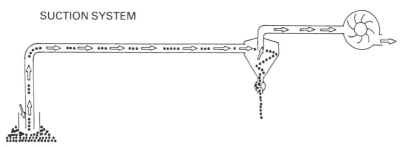

Conveyance from several intake points to a fixed outlet is the main advantage of a suction system.

BLOWING SYSTEM

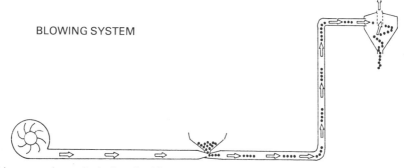

This system is particularly suited for conveying from a fixed intake point to several outlets.

Fig. 2.5 Pneumatic conveying systems (*Kongskilde*).

and each kilowatt of power input will convey only about 1 tonne/h over a distance of 20 m. Furthermore, noise and dust are particular problems.

Material may be conveyed up to 150 m and for such an output the conveyor would normally be tractor mounted, needing a minimum of 40 kW pto power. Increasing the conveying distance from 10 to 100 m will reduce throughput to about one-third. An excessive number of bends also reduces output.

The conveying system comprises a centrifugal fan which produces high speed air flow at 18–23 m/s at about 4 m water gauge pressure, feed entry or exit valves, a cyclone to separate the material from air and tubing of 140–160 mm in diameter. However, a version of pneumatic conveyor developed in the USA uses small bore pipes to deliver meal over long distances. An auger introduces meal into an air stream produced by a compressor with delivery through 50 mm bore pipework.

The three common arrangements used for pneumatic conveyor are depicted in Fig. 2.5. In the blowing system the conveying pipe is connected to the pressure side of the fan which provides a powerful current of air. Material to be conveyed enters the system through a feeder in the form of venturi or rotary valve. A discharge cyclone is placed at the end of the line to permit separation of material from the air. In the suction system the conveying pipe is connected to the suction side of the fan. Material is sucked into the free end of the pipe by means of a venturi, rotary valve or suction head. Conveyance from several places of intake to a single location of outlet can be organised by using diverter valves. At the base of the cyclone, material leaves the system via a rotary valve which prevents air being sucked in. The suction system tends to be less dust prone than the blower system.

The suction/blowing system gives a good deal of versatility as the same fan can be used for both systems. Grain by-passes the blower by being sucked into a cyclone where air and grain are separated – the air being drawn into the fan and the grain reintroduced into the air stream via a rotary valve. Ancillary equipment, such as a weigher, can be introduced and it is fitted between suction and blowing sections. Transportable self contained suction blowers are available with which grain can be sucked up from floor storage and blown some distance to the feed processing plant.

Bucket elevators

Bucket elevators are the obvious choice for vertical conveying, particularly where high vertical lifts are required and where feed might easily be damaged, particularly in cube and pellet form. They are capable of elevating efficiently to 18 m at up to 120 tonnes/h if required with a low specific power requirement, which might be around 4.1 kW for a 50 tonnes/h capacity. These elevators comprise a vertical, often wooden box with pulleys top and bottom around which an endless rubberised belt is driven with bucket shaped metal scoops attached at regular intervals. The elevator is either fitted with an intake hopper at the base when, for instance, it might be used in association with a cuber, or more commonly it is set into a below ground material reception pit.

Material picked up at the base entering on the front of the elevator, that is the rising side of the belt, is elevated and discharged after passing over the top pulley at the back of the elevator. Back feeding is possible but this results in slightly lower output. Drive is provided by an electric motor via a V-belt to the top pulley. Twin leg versions are available where a single drive unit drives twin elevators giving opportunity for two separate materials to be conveyed simultaneously, each within its own casing.

Output depends upon size of motor fitted, capacity of each scoop, scoop spacing and belt speed. Belt speeds of 1.27–1.53 m/s are used, at higher speeds discharge becomes inefficient.

Most designs incorporate means of cleaning out the elevator boot and tensioning the conveyor belt. Rotation detectors can be fitted to monitor shaft speeds and give an audible warning of belt slip.

Ingredient measuring systems

As concentrate feed constitutes such a great proportion of the costs of livestock production an efficient means of measuring feed is an essential tool in the proper management of any livestock enterprise. Where feed is bought in no special bulk measuring facilities need be provided and feed can be measured via feed dispensing rationing devices. However, for on-the-farm feed processing plant every endeavour should be made not only to ensure that all ingredients going into the mix are measured with reasonable accuracy – particularly those which form a small proportion of the ration – but also to keep a check on quantities of feed mixed. Investment in a realistically accurate feed measuring device can pay handsome dividends and may also provide facility for automatic dispensing of ingredients in a fully or semi-automated plant.

Feed can be measured in two ways: on a volume basis or by weighing. Volumetric systems are generally cheap but prone to inaccuracy due mainly to variations in bulk density and moisture content between batches of the same materials, while in recent times there is considerable interest in weighing systems because of their inherent accuracy.

In the following section various means of measuring bulk quantities of raw materials or mixed feed will be discussed. However, the reader should refer to the section on mixing in Chapter 1 for fuller details of proportioning devices used with certain feed processing plant. Similarly, within individual chapters on cattle, pig and poultry feeding, details are given of measurement of feed for rationing purposes.

Volumetric measurement

In general terms, there are two systems of volumetric measurement: batch volume and calibrated conveyors.

Batch volume systems are based on calibrating and suitably marking either mobile containers or static holding hoppers. Thus, in the simplest case plastic bins, barrows or feed trucks, when carefully calibrated and clearly marked, provide a ready cheap means of measuring feedingstuffs on a small scale. Accuracy is influenced by whether feed is levelled by the stockperson, hopper shape and frequency of recalibration. For larger quantities bins can be fitted with a long transparent plastic window with calibration marks to indicate the volume being held. Here calibration should take into account the angle of repose which will, of course, vary with moisture content. Such systems are prone to considerable inaccuracy, not least to inability to see material clearly.

Calibrated conveyors provide perhaps a better potential means of volumetric measurement. Simply, any suitable conveyor, such as an auger, is calibrated by timing how long it takes to fill a bucket with feed and weighing the contents of the bucket to determine throughput. Then, by timed manual switching, or using an automatic time switch system, an appropriate quantity of material can be delivered to, say, a common hopper above a mill or mixer. A number of such conveyors can be arranged to deliver automatically, in sequence, appropriate amounts of each ingredient to the feed processing plant. Calibration of augers can be organised on either the basis of time or number of revolutions. Provided auger flights are kept full then accuracy can be in the region of $\pm 2\%$, but changes in bulk density and moisture content, grain impurities and malfunction of the conveyor (for instance slipping belts) do have a detrimental effect upon accuracy. It pays to recalibrate frequently, particularly after new batches of materials have been delivered.

Gravimetric measurement

There are increasingly well recognised severe potential penalties arising from inaccurate measurement of livestock feed whether at the feed processing stage, which can affect feed cost and quality, or in rationing feed to stock, both of which influence livestock performance and profitability. Thus, in recent years, the use of feed weighers has become progressively more widespread with a wide range of equipment from which to choose.

The potential cost of any farm feed weighing system depends on the degree of accuracy acceptable, the sophistication of automation associated with it and installation including any attendant conveyors. Some weighers are available stamped to DTI standards and, depending on the weigher, may be accurate to within $\pm 0.25\%$ or in some cases 0.1% of full scale deflection. While these weighers can be used for trading, they are somewhat expensive and it is not normally necessary for such fine degrees of accuracy for farm concentrate feed. In basic format a weigher gives a visual reading of quantity on a scale, but usually some form of counting and automatic control is integrated to provide a series of automatic weighings and also the capability of operating associated conveyors. It is often quite difficult to add a weigher to an existing feed handling system – headroom requirement being the common limitations – so that it is best, if possible, to build a weigher into the overall system at the outset. Additional or extended conveying equipment will commonly be needed to serve the weigher, and the total installation cost will often exceed the cost of the weigher itself.

Weighing can be organised on a batch or a continuous basis.

BATCH WEIGHERS

Batch weighers are split into two main categories: those suitable for mobile bulk consignments, such as weight bridges and weigh pads, which can accept trailers or lorries of bulk material such as grain, and static batch weighers, such as platform scales and weigh hoppers. The latter category is the most appropriate for concentrate feed batch weighing.

Platform scales

These are widely used for weighing individual sacks and are capable of weighing from, say, 0.25–0.5 kg to several hundred kilograms. Most farm feed processing plant need such scales for weighing small quantities of ingredients such as minerals and vitamins and also perhaps for weighing small quantities of rations in sacks. Three main types are in common use, each with a simple platform on which sacks or other containers are placed. By a series of linkages and cranked levers connecting from the platform, weight is registered either on a rotating circular scale with needle or via a manually adjusted yard arm. The latter comprises a lever system mounted on a knife edge where a number of small weights are added to a hanger at

Plate 2.4 Automated beam feed scale (*Big Dutchman*).

the end of the balance arm and fine measurement is achieved by moving a weight along a graduated 'poise' scale to balance the load on the platform. A typical capacity for this type of weigher is 250 kg with the poise scale marked 10 kg × 0.1 kg. Increasingly today, platforms are mounted on load cells with the weight registered on a digital readout.

One particular form of yard arm platform scale has been adapted to accommodate the supply hopper of a mechanised chain feeding system for poultry. Once the amount to be dispensed (0–725 kg) is set on the yard arm, the supply auger fills the hopper until the scale microswitch shuts it off.

Where significant quantities of rationed feed requires to be bagged up, for instance for an outdoor pig herd, an electronic sack weigher/filler may be appropriate. The operator secures the sack on to the weigher using a quick action clamp and opens the feed gate. This is held open mechanically until the sack is nearly full, then a trip operates to close the feed gate partially until the final accurate weight is reached, whence a further trip cuts off the feed and closes the gate. Up to 150 sacks an hour can be filled in this way, but the operator has to work quite hard!

Weighing hoppers

These are a common feature of a number of on-the-farm batch feed processing systems. Feed is delivered, generally by auger, to a tapering hopper with a capacity of 0.5–5 tonnes, that is, enough for one batch of feed. The hopper contents are weighed either mechanically with a yard arm or spring balance or more commonly using load cells. Materials are added in sequence and the system can be automated by the incorporation of microswitches which can be pre-set to control particular augers filling the hopper.

With the mechanised weigher the hopper may be suspended or supported in a cradle fitted with a spring balance or simple graduated yard arm which may be moved progressively by hand to the required setting to achieve the cumulative weight of the various ingredients. Microswitches and relays can be added to automate the switching of augers.

Where height permits, the hopper may be installed on legs at high level above a roller/hammer mill which, in turn, is mounted on a chain and slat mixer, thus exploiting gravity for handling. Alternatively, the hopper may be suspended just above floor level with an emptying auger running the length of the base of the tapered hopper – a useful system for handling meal and additives from bulk storage bins. Again, loading augers can be controlled manually or automatically in sequence. In another version, the weigher consists of two hoppers, one mounted above the other. The higher weighing hopper is fitted with a 'bomb door' to provide quick emptying to the lower hopper which serves as the mill reserve hopper. This provides semi-continuous weighing without interruption of the milling process.

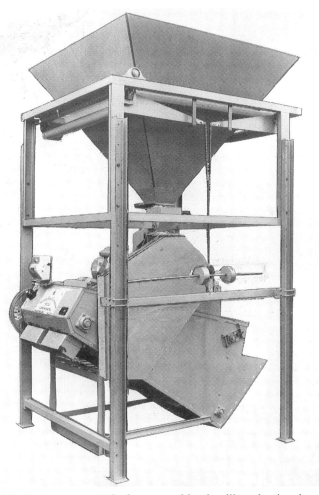

Plate 2.5 Weigh hopper mounted above combined mill and mix plant (*Alvan Blanch*).

In many ways load cells provide a more convenient means of weighing and they are becoming very widely adopted, not just in association with feed processing but also, for example, for liquid feeding of pigs. Here the hopper is fully or partially mounted on hydraulic or electronic load cells which, in some instances, can be retrofitted to an existing hopper. With hydraulic load cells the load is applied to a liquid filled system which measures, by means of a dial gauge, the hydraulic pressure generated when the load is applied to a hydraulic cylinder. By fitting adjustable pointers to the scale linked to electric controls, augers can be started and stopped to feed material into the weigh hopper.

Electronic strain gauge load cells are being adopted more commonly today. The strain gauge, comprising a filament of very thin wires, is bonded

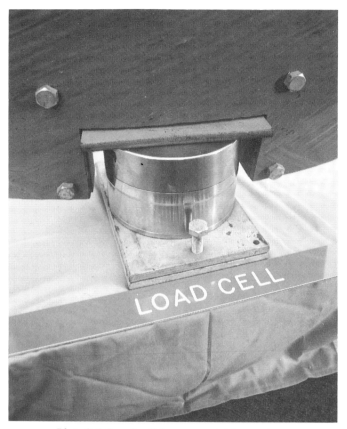

Plate 2.6 Load cell hopper weighing system.

to a steel bar which is subjected to a bending moment by the weight to be determined. A small electric current is passed through the wires whose resistance changes with the stretching or contraction produced by the bending of the bar. The change in reading is measured and calibrated to give a direct readout of weight. Strain gauge load cells are relatively robust, can be temperature compensated, are very accurate and are easily incorporated into an automated system.

One computerised weighing system comprises a weigh hopper mounted on load cells and acts as a 'control centre' for larger mill and mix plant. Batches of ingredients are weighed in sequence and passed to the mixer. The programmable controller can control the whole operation and it is therefore a means of improving the accuracy and control of an existing feed preparation plant. The system has the capacity to store up to ten ration recipes with up to ten ingredients. In a recent version the weigh hopper now incorporates two vertical mixing augers so providing an integrated weighing and mixing system.

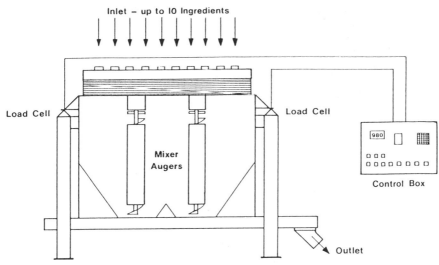

Fig. 2.6 Diagrammatic view of BB Weighmix.

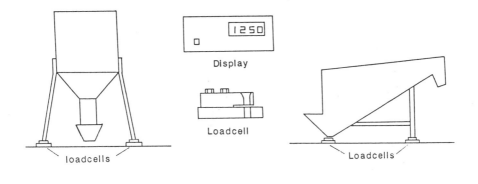

Fig. 2.7 Load cell mounting for mixers.

Electronic load cell kits and control systems are available for fitting into existing mill and mix plant and feeding systems. Typically, feed is weighed either by mounting the mixer on three load cells as shown in Fig. 2.7 or by a kit comprising a flexible 'hopper' suspended from a load cell, as shown in Fig. 2.8. Such kits can be supplied either with digital readout for manual control of conveyors or with controllers for automatic selection of ingredients. The controllers range from a basic system incorporating two pre-set ingredient weights up to a sophisticated system providing pre-programming of up to fifteen rations, each of up to fifteen ingredients.

Where trailers of grain or mixed feed need to be weighed and there is

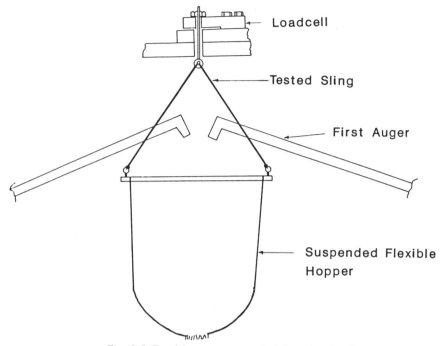

Fig. 2.8 Feed hopper suspended from load cell.

no nearby weigh bridge, the use of compact, portable hydraulic or strain gauge load cell pads is a possibility. Two or more platforms are placed on the ground and the vehicle driven over them until two or more wheels rest on the weigh pads. Accuracy of around ±3% for hydraulic weigh pads is often limited by the capacity of the operator to interpret the markings, which are necessarily small on these portable weighers.

CONTINUOUS WEIGHERS

On-line continuous weighers do offer a greater degree of flexibility compared to batch weighers and there is a particularly wide range available. In principle, many continuous weighers comprise, in effect, a small batch weigher organised to fill and discharge repeatedly when installed within the material flow line. Some are considered to be very accurate, but this property is a function of the weighing principle and shut-off device as well as the fill capacity. In general terms, greater accuracies are often achieved with a smaller number of large consignments.

The most accurate continuous weighers have a 25–100 kg capacity single hopper supported on a counter balanced beam arm. Solenoid operated shutters regulate flow into and out of the hopper controlled by sensitive microswitches actuated by the tip of the weighing lever which detects various degrees of balance. The system is arranged to reduce the high initial flow of material to a trickle for the last few seconds of the weighing period, which is followed by a very rapid cut-off when the true point of balance is reached. Batches can be varied typically from 10–100 kg.

One of the simplest continuous weighers is operated on an oscillating principle comprising a small, simple split hopper sited under a material outlet. The flow oscillates the hopper as it fills and empties and this rocking operates a counter. Automatic control is possible using a pre-set counter.

Counterbalanced tilting weighers operate on a similar principle – but are more accurate – with a hopper capacity of 10–50 kg and throughput of 6–20 tonnes/h. Such a weigher comprises two siamesed type symmetrical hoppers, with steeply tapered base to ensure quick and complete emptying, which are pivoted about a horizontal axis in the plane of the partition. The split container is supported on a forked lever system incorporating a counterweight bar at one side. The hopper is prevented from tipping sideways by a simple pin engaging in a slot at the top of the hopper. When the hopper becomes full it balances the counterweight and the container

Plate 2.7 Counterbalanced tilting weigher.

moves downwards. The beginning of this movement reduces the incoming flow and the final part of the movement unlocks the securing latch allowing the hopper to tip sideways. At the same time flaps at the base of this section of the hopper are opened automatically permitting the contents to empty, and while this is happening the other part of the hopper can now be filled. The process is repeated, each time operating a counter which logs the amount of material passing through. Automatic switching can be incorporated to stop the weigher when a predetermining weight has been reached.

Another mechanical type of continuous weigher comprises a rotating paddle mounted in a horizontal cylinder, into which grain is introduced at the top. The paddle is mounted on a central shaft and incorporates a cam plate at one end which makes contact with a fixed stop pin. The paddle is normally loaded into a near horizontal position by the pin preventing the grain from passing through the cylinder. The paddle shaft is mounted in a pivoted cradle, the other end of which includes an adjustable counterweight. When sufficient grain has collected on the lower end of the paddle to overcome the counterweight, the paddle sinks slightly and the cam plate will clear the stop permitting the paddle to rotate through 180° and release the batch of grain which drops through the bottom of the cylinder. Each successive rotation is registered on a counter, and an automatic pre-selection control panel can be incorporated for autoweighing of several ingredients according to programmed quantity. The weigher has a capacity of up to 800×10 kg weighings per hour.

There are two main types of electronic continuous weigher. The continuous belt weigher comprises a short belt conveyor pivoted at its lower end and supported at the higher end by a load measuring device such as a spring linkage, but more likely a load transducer employing a strain gauge. Material is delivered along the belt and a downward force is exerted on the load transducer in proportion to the weight carried. This force is integrated with belt speed and the resultant reading transferred to an electronically operated counter. Facilities can be incorporated to stop the belt when a certain quantity of material has been delivered. Accuracies up to $\pm 2\%$ or so are possible and the major advantages of this type of weigher are its capacity for very high outputs of up to 30 tonnes/h and its capability of handling virtually all feeds, including silage. However, it is not so accurate at low throughputs as might be typically experienced in handling concentrate feed.

A newer type of electronic continuous weigher comprises a drum suspended on load cells with a geared motor drive to rotate the drum and so empty the contents. The drum is mounted in a water- and dust-proof container – a welcome feature as many weighers are prone to produce clouds of dust. Feed enters through a supply hole at the top and leaves at the bottom, usually by auger. The unit also incorporates a microcomputer and digital display. The weighing sequence comprises weighing of the

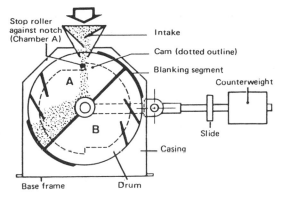

Grain entering weigher (Chamber A) from intake. Drum is held stationary by stop being engaged with cam notch.

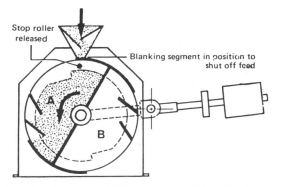

Grain weight in Chamber A has overridden counterweight, allowing drum to drop and blanking segment to move round and shut off supply from intake.

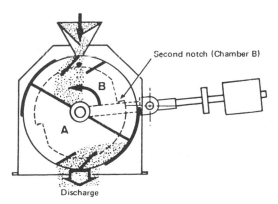

Drum has moved further round. Chamber A is discharging. Chamber B is filling.

Fig. 2.9 Rotating continuous weigher (*Law-Denis*).

Plate 2.8 Electronic feed weigher incorporating a rotating drum suspended on load cells (*DOL 99*).

empty drum followed by starting of the filling auger. The filling auger is stopped when the set level (between 10 and 30 kg) is achieved, following which the drum is weighed. The geared motor now inverts the drum, the drum is weighed again and the net weight of material is added to the total counter. The unit has a capacity of up to 3700 kg/h and can handle grain, meal, pellets or cubes. A restrictive feeding facility over a specific time period can be obtained by pre-setting a time clock as required. During the feeding period the weigher can dispense from 5 kg to over 9000 kg.

Yet another type of electronic continuous weigher is available but as yet is confined to estimating yield of grain delivered in the field from the combine harvester discharge auger. These electronic impact flow meters comprise an inclined plate on to which the grain flows. With careful design, the vertical force on the plate can be made proportional to the flow of grain and weighing accuracies of 2–3% can be attained. In the future such weighers may find a role in the weighing of animal feed.

References

Electricity Council *Grain Drying and Storage*. Farm Electric Handbook.
McLean, K.A. (1980) *Drying and Storing Combinable Crops*. Farming Press.

3 Mechanised pig feeding

As pig units have significantly increased in size, manual systems of feed transporting and dispensing have come under close scrutiny. More effective use of pigmen's time for stockmanship and the reduction of stress to both the pigs and pigmen have proved compelling reasons for the steady growth in mechanised pig feeding systems. The recent availability of computer controlled automated feeding has further enhanced the potential for more accurate feeding for both finishing pigs and dry sows – hence the scope for improvement in herd performance.

As far as finishing pigs are concerned, the Meat and Livestock Commission (MLC) figures for current UK feeding systems are shown in Table 3.1. In general terms, dry feeding systems have been favoured in the east and north of the UK, whereas pipeline feeding, often dependent upon availability of milk by-products, is more popular in the south. Modernisation and investment in plant has been blighted to some extent among pig herds in recent years as a result of the extremely volatile pig trade. But overheads are spread more easily with increasing size of unit, and for larger herds, some form of mechanised feeding really becomes a necessity.

Automated and semi automated pig feeding – the benefits and disadvantages

Systems are now available to automate pig feeding completely from the raw material stage to delivery and dispensing of an exact amount to any particular pen of pigs at pre-set intervals. But equipment is available in less sophisticated systems at least to improve on manual feeding. So what are the particular benefits of automation?

Benefits

(1) REDUCTION OF TIME AND EFFORT
Although labour constitutes only a relatively small proportion of the total cost of pig production (around 9%) for birth to slaughter heads, 7% for

Table 3.1   MLC figures for current UK fattening pig feeding systems.

	Year to Sept 87	Year to Sept 86	Year to Sept 85
By hand	46	41	41
Wet feed by pipeline	26	28	28
By automatic or semi automatic system	15	15	15
By hopper	13	16	16

fattening only), as a rule of thumb automation gives scope at least to double the amount of stock that one person can look after. But much depends on the degree of automation involved. Thus, a survey on pipeline feeding showed an average of 2.6 man minutes per 100 pigs spent feeding, ranging from 4.7 man minutes per 100 pigs using manual valves to 1.5 man minutes using automatic valves. With electronic sow feeders, labour saving can be more apparent than real, as time saved in auto dispensing of feed has to be devoted to more time checking sows, refitting collars, training sows, and so on.

(2) SCOPE FOR ENHANCED STOCKMANSHIP
Freeing the stockperson from the drudgery of carting food around, and possibly dispensing it, should provide him or her with better opportunity to observe stock. The returns on good husbandry can be high. But there is the potential danger that complete automation of the feeding process can promote bad habits and hence many producers prefer semi automatic systems where the stockperson has to be present to dispense feed to individual pens and hence keep an eye on stock behaviour.

(3) REDUCED STRESS
Automatic or semi automatic feeding significantly reduces feeding time and this should reduce stress on both pigs and pigmen. Noise levels of up to 100 dBa have been recorded in some piggeries – a definite safety hazard for stockpersons after prolonged exposure! Ad lib and simultaneous drop feeders also give scope to curtail feedtime excitement and agitation, and here pigs (and stockpersons) should expend less nervous energy! But this is difficult to evaluate in cash terms.

(4) POTENTIAL FOR INCREASED FEEDING ACCURACY
With feed comprising 80–85% of total costs of producing the fattened pig, any improvement in feed conversation efficiency, and also daily liveweight gain and grading, is important.

Automatic dispensers, particularly those operating on a weight rather than volume basis, can facilitate very accurate rationing and hence control on feed intake. But contemporary automatic feeding systems have been somewhat revolutionised by load cell and microchip technology, giving potentially very accurate feed control. Not only can such systems accurately control the amount being dispensed but they also offer sophisticated control over feeding regimes, including automatic adjustments of rations.

(5) GREATER FACILITY FOR USE OF NON-CEREAL SUBSTITUTES

Entrepreneurial buying of food by-products can make a significant contribution to the reduction of overall feed costs. Mechanised systems, such as pipeline feeding, can most readily deal with such products as whey, skim milk, brewers waste, potatoes and other materials, whether in liquid or solid form. Some computer control systems are capable of compiling the most cost effective ration from a wide variety of ingredients.

Disadvantages

(1) CAPITAL OUTLAY

Even less sophisticated systems will increase capital investment and this will generate an ongoing depreciation cost against the cost of each pig produced.

(2) MAINTENANCE/SPARES

Equipment will need to be kept in good mechanical and electrical condition, with a reliable source of spares and service backup.

(3) RELIABILITY

Proven reliability should be a vital feature in the selection of equipment. For operation 365 days a year not only should manual override be built in but, ideally, speedy correction of faults should be possible. Continuity of electrical supply is, of course, vital and a stand-by generator should really be an essential tool. The feeding system should be fail-safe and effective alarms should be built in to protect the welfare and performance of stock.

(4) FITTING INTO EXISTING BUILDINGS

Some feed systems are difficult to integrate into existing buildings. Where building adaptation is not possible, investment in a new one may be necessary. On the other hand, electronic sow feeding has facilitated the use of perhaps previously unsuitable buildings (such as large straw yards).

(5) LIMITATION OF FEEDING DIFFERENT DIETS IN ONE BUILDING

Most mechanical systems do not accommodate the opportunity to vary types of ration between pens of pigs in the same building.

(6) ENCOURAGEMENT OF UNDESIRABLE ANIMAL BEHAVIOUR

Some mechanical feeding systems can promote undesirable behaviour such as bullying, dunging in the lying area, vulva biting in sows and non-acceptance of the feeding system (for example, with electronic sow feeding some sows refuse to enter the feed station). A few exploratory visits to other producers to learn of their experiences are essential before launching into a new system. Some equipment companies also offer group 'think tank' sessions when such problems can be discussed and improvements promoted. Care in equipment design and installation and due thought given to promotion of the 'correct' pig environment (for example, pen layout, aerial temperature and air flow pattern) can significantly reduce such problems.

(7) PRODUCTION OF DUST AND WASTE

Some systems, particularly those using floor feeding and meal, are prone to produce dust and encourage fouling and waste of feed.

Further considerations

Short term feed storage

Provision needs to be made for sufficient short term storage capacity of feed, in whatever form, to give sufficient buffer supplies against holidays, weekends and possible lack of deliveries in inclement weather. At the very most, feed may need to be stored for up to 8 weeks without deterioration, but it is more usual to expect to carry stocks for 10 days to 3 weeks or so.

As a guide to quantities of feed to be stored, Table 3.2 gives a rough estimate of typical feed consumptions. It should be borne in mind that the true average weight of pigs in a house will determine the average daily requirement. If the average weight at any one time, however, is somewhat greater than the arithmetic average of the youngest and oldest of groups then significantly more feed will be needed.

EXAMPLE

To determine the minimum short term storage need for bought-in feed for a fattening house holding 320 baconers, taking the specific volume of meal/small pellets at 1.73 m^3/tonne and 20 tonne load deliveries:

at an FCR of 2.5   800 kg/day will be consumed on average
while at an FCR of 3.0   960 kg/day will be consumed on average

A 20 tonne load will therefore last 25 or 21 days respectively. In practice, capacity for a few tonnes overlap is necessary for at least a further 3 days' supply to allow for weekends/Bank holidays. This gives a minimum storage need of 22.4 tonnes and 22.8 tonnes respectively. It would be prudent

Table 3.2 Typical feed consumptions.

	Wt gain assumed	Assumed period (days)	FCR	Av daily consumption + 20% (kg)
Porkers	45	100	1.9	1.03
			2.4	1.30
Cutters	66	115	2.3	1.58
			2.8	1.92
Baconers	75	135	2.5	1.66
			3.0	2.00
Heavy Hogs	100	165	3.0	2.18
			3.5	2.55
Sows	Approximately 1320 kg/year Allow average maximum 3.6 kg/day per sow, this is generous and allows for extra feed needed by suckling sows.			

to allow for an extra 10%, so a 25 tonne bin would be suitable in this case. Taking the specific volume of meal or small pellets at 1.73 m³/tonne this requires a bin of 43.25 m³ capacity (for example, 2.85 m diameter and a total height of 10.4 m).

Bulk feed bin features, including means to minimise bridging, are given in Chapter 2.

For storage of ingredients other than meal and pellets (for example, milk by-products and brewers waste) particular attention should be paid to storage shelf life, particularly in warm weather. Enough space should be available to store economic consignments. Hygiene is another important consideration and containers should be capable of being readily cleaned and sterilised. In addition, the particular material's flow characteristics will determine the appropriate type of filling and emptying facilities.

Suitability of conveyors

For conveying meal it is vital that conveyors should in no way 'unmix' feed by separation of coarser and finer ingredients. Low pressure/high volume pneumatic conveyors are particularly prone to this problem. Dust is a further hazard of pneumatic conveying.

Where pellets are to be delivered, most conveyors, including high pressure/low volume pneumatic conveyors, are suitable but augers will break up pellets unless they are slowed down to about half speed or unless they are of the centreless type. Details of suitable conveying systems are given in Chapter 2.

When wet feed is conveyed, the likelihood of blockages will be influenced by grist size, pipe diameter, dilution rate and pump type and capacity. In general terms, grist larger than 3 mm in size and mains pipework less than 50 mm in diameter should be avoided, as should viscous solutions of less than 1:2.5. Centrifugal pumps are normally used but helical rotor pumps are capable of handling viscous solutions. These are more fully discussed in the later section on pipeline feeding.

Degree of automation

In simple terms, any mechanical feeding system might incorporate one of three main forms of control:

(1) SEMI AUTOMATIC CONTROL
Here feed is typically conveyed mechanically and dispensed manually. Alternatively, the operator fills containers which are mechanically dispensed later. This is a lower cost option, still fairly labour intensive, although it can take much of the chore out of moving feed and gives the stockman on-the-spot control and responsibility of amounts fed.

(2) SIMPLE AUTOMATIC CONTROL
Here, conveying and dispensing is undertaken automatically, usually with a simple timed device to trip the feeding system and refill it. For pipeline feeding, filling a batch mixer might still be controlled manually and, once mixed, feed is delivered and dispensed automatically.

(3) SOPHISTICATED AUTOMATIC CONTROL
Powerful computer controllers are now widely available to provide enhanced rationing control with a range of capabilities. Typically linked to load cell weighing of wet mixers, these computer controllers can not only monitor amounts of individual ingredients mixed, with a facility to be programmed to mix a number of ration formulations, but also, fundamentally, they are capable of very accurate control of amounts fed to each pen. Most systems allow opportunity for feeding little and often, sometimes regarded as the most efficient means of using feed. Several systems may be programmed to adjust incrementally feed scales and hence follow pigs' 'growth curves', allowing them (one hopes) to achieve their full potential. Other facilities include total information on feed usage and costs: this ensures that management is provided with immediate and up-to-date confirmation that desired feeding scales are being followed – a vital element in effective enterprise control. By linking such a control to electronic pig weigh scales, it is possible to build up a comprehensive analysis of physical and economic performance.

At least one such current control system provides a portable data input module so that adjustments to feed inputs can be made, if need be whilst observing any pen of pigs. Later, when the hand module is plugged into the main computer, data is downloaded from it.

Electronic dry sow feeders are perhaps the most well reported example of such computer controllers. Automatic recognition of individual sows, group housed, enables the computer to dispense feed as appropriate to each and compile, store and analyse a range of useful performance indicators. This is discussed in more detail in a later section.

Mechanised feeding systems for fattening pigs

Table 3.3 summarises the main systems of mechanised feeding of fattening pigs.

Factors affecting choice of system

(1) ACCURACY

Complete accuracy obviously cannot be expected, but it is generally accepted that any system should be capable of dispensing feed to within ±5% on a day-to-day basis. The fact that bulk feed density of similar feed can vary by up to 10% for consecutive batches does mean that volumetric dispensing is certainly prone to a certain degree of inaccuracy. Regular and thorough calibration will certainly help to minimise these effects. There is a clear trend towards gravimetric dispensing systems using either weight operated dry dispensers or load cell wet feeders. Such systems should certainly be capable of accuracy to within ±5% or better, and Agricultural Development and Advisory Service (ADAS) field surveys have confirmed that some computerised liquid feeding systems are well within this figure. With many feeders, errors tend to be greater the smaller the amount dispensed at one 'drop'.

(2) ANIMAL PERFORMANCE

Several well documented experiments have compared the performance of fattening pigs fed on meal, cubes or liquid feed. In general terms, cubes and liquid feed give significantly better FCRs, although growth rate is generally faster with pellets and sometimes is marginally better with dry rather than wet feed. On-floor feeding does create more dust and waste by fouling, hence wet feeding mostly has the edge as far as FCR is concerned. Floor feeding with simultaneous feeders does reduce stress, but there is little evidence to translate this into financial terms.

Ad lib feeding is not universally acceptable. Pigs from some genetic sources are not suited to the system, resulting in poorer carcase grading than with restricted feeding. Having said that, with current strains of pig available there has been a resurgence of interest in ad lib feeding in well insulated, environmentally controlled fattening houses.

Table 3.3   Summary of current mechanised feeding systems available for fattening pigs.

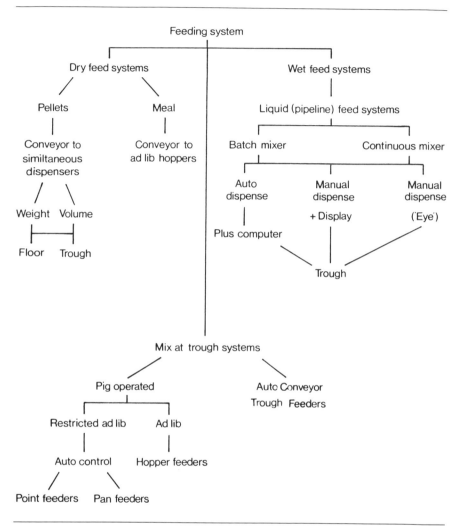

(3) FLEXIBILITY OF SYSTEM

The facility to vary feed formulations readily is an important aspect for many producers and this is where pipeline feeding especially comes into its own. Opportunity buying of food processing byproducts can certainly make producers less reliant on traditional feed sources and gives scope for trimming feed costs. Pipeline feeding also facilitates the opportunity for medicaments to be quickly incorporated and fed when required. In addition, batch pipeline feeding integrates particularly well with on-the-farm milling and mixing systems with the possibility of forgoing the need for a dry feed mixer.

(4) CAPITAL AND RUNNING COSTS

Capital costs are influenced by the specific requirements of individual circumstances, including the need for building adaptation and degree of automation, but most importantly by the scale of operation. Considerable economies are possible in the cost per pig place for larger installations. As a rough guide, the following summary gives typical installation costs. Figures do not include storage bins, or tanks nor troughs.

- Dry feed dispensers approximately £9–£14 per pig place
- Pipeline (manual valves) approximately £11–£16 per pig place
- Pipeline (auto valves) approximately £19–£24 per pig place

Ignoring labour costs, electricity represents the main running cost, together with spares and maintenance. Electrical consumption of the order of 0.15–0.25 kWh/100 pigs per day will be likely for dry feeding and 0.3–0.8 kWh/100 pigs per day for pipeline feeding.

Simultaneous dry feeding systems

In general, most of these systems dispense feed on to the floor, but it is equally possible to dry feed into troughs using, for instance, a linear dispenser. The most noticable impact of simultaneous dispensing is in minimising pig excitement and reduction of noise at feed time.

On-the-floor feeding is not suitable for fully slatted fattening pens and meal is generally unacceptable, thus in order to minimise the inevitable dust problem pellets must be fed and, for home-mixers, the extra cost of pellets has to be accepted. Inevitably some feed is wasted by fouling and treading into dung which means it is difficult to achieve an optimum FCR. Despite these disadvantages, floor feeding remains a very popular system, influenced no doubt by low capital requirements and lower building costs not only due to lack of troughs but also because it is possible to house more pigs in the space saved with flexibility of pen shape.

Figure 3.1 shows the basic layout of a typical on-the-floor feeding system. The system comprises a feed storage bin located outside the fattening house, a feed conveying system, hoppers with adjustable rationing control for each pen, together with an automatic dispensing system. Many systems load the conveyor directly from the storage bin but some incorporate a separate conveyor from the base of the bin, filling a separate loading hopper which supplies the main conveyor. The latter is either of the centreless auger type or endless chain/cable and disc type, although other conveyors can be used. The maximum length of single conveyor will be limited to about 65 m for centreless auger, but up to 250 m is possible for chain and disc types. Feed is conveyed overhead around the piggery

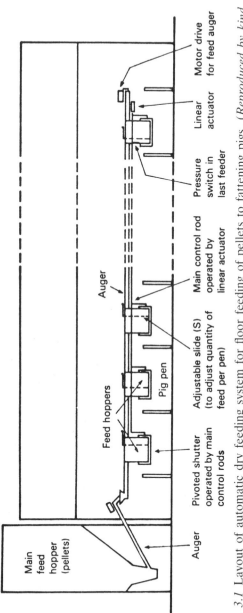

Fig. 3.1 Layout of automatic dry feeding system for floor feeding of pellets to fattening pigs. (*Reproduced by kind permission of Farm Electric Centre (Electricity Council)*.)

supplying feed hoppers, with typical capacity of 18–36 kg, suspended from the conveyor tube or from the roof, one sited over the centre of each pen. When the first hopper has been filled with the appropriate amount, feed is supplied to the next until a pressure switch at the last feeder shuts off the conveyor. At feeding time the dispensing system is tripped either manually or, more usually, automatically, to release feed simultaneously from all hoppers. Hoppers are immediatcly filled whilst pigs are feeding in order to minimise 'anticipation excitement' at feed time. Dispensing can be achieved in a number of ways employing, most popularly, a common cable and winch system, with attachments to individual hopper release devices, or a rod type linear actuator.

Typical hopper release arrangements include:

- Pulling apart a pivotted single or double shutter normally held closed by spring(s) (see Plate 3.1).
- Raising a conical plastic release bung from the floor of a narrow hopper by cable release or rotating rod.
- Raising of the cylindrical hopper feed container off a fixed cone shaped base, releasing feed in a circular pattern approximately 1.5 m in diameter (see Plate 3.2).
- Rotation of a release plate using a linear actuator.

Plate 3.1 Weigh type dump feeder (*Collinson*).

Plate 3.2 Volumetric feed dispenser. Feed is released when the cylindrical feed container is raised, by cable winch, off the fixed cone shaped base (*EB*).

Plate 3.3 Trough feeder (*Funki*).

In another version, cables and rods are dispensed with by the use of a counterbalanced bottom hinged flap and solenoid operated release catch. Feed is measured into hoppers either volumetrically or by weight. Volu-

Plate 3.4 Volumetric dump feeder (*Collinson*).

metric control is most commonly achieved by adjustment of either a calibrated telescopic drop or a slide (see Plate 3.3). The higher the free outlet point of the discharge tube from the base of the hopper, the more feed delivered. In another design, a feed retaining plate, with trapdoor, is raised or lowered within a narrow tube hopper on a central rod with screw thread. With another novel design the centreless auger conveyor passes through a long compartmented hopper with an outlet to each compartment. Each outlet may be closed up or opened by rotating a simple rotary slide. The volume can be adjusted to give ten divisions, each of 2 kg.

There is a trend, however, towards weight operated rationing devices, which are more accurate. Each hopper is suspended from a pivoted counterbalanced weigh beam which incorporates a cut-off flap. The weigh beam, extending to one side, is calibrated to enable a moveable weight to be set and locked at a point which will exactly counterbalance the required amount of feed delivered. Once this has been delivered the weigh beam lifts, tipping a pivotted flap to shut off the feed supply to the hopper.

Versions of simultaneous dry feeders are also available to distribute to troughs or circular feeders. Included amongst these is a scaled down version of the point dispensers used for floor feeding (see Plate 3.3) which can also be used for feeding individual stalls, for instance dry sows. Other types incorporate a series of telescopic drop tubes aligned above the trough and operated together giving a line of feed. Alternatively, a linear dispenser provides the same result. The latter, sited immediately above

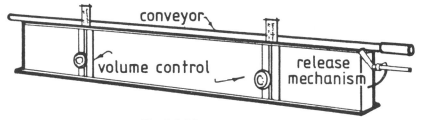

Fig. 3.2 Linear dispenser.

the trough, consists of an elongated rectangular hopper with feed conveyor mounted above and a series of telescopic drop tubes. Feed is released from the hopper to the trough by means of a longitudinal hinged bottom flap.

For all these feeders the main aspects of good design include accuracy over a range of quantities delivered and ease of adjustment. A particularly important condition is an adequate means of preventing any one hopper from filling with feed when a pen is empty, so as to avoid waste and deterioration of feed. This is quite easy to achieve with weight operated feeders when the counterweight can be moved to the zero position. With drop tube versions a cut-off slide at the top of the tube can prevent the build-up of a column of stale feed. Flexibility of control should permit at least two automatic feedings every 24 hours, with automatic refill and alarm facility.

Ad lib feeders

Ad lib feeding is commonly used for growing pigs put into fully slated pens after weaning. Until relatively recently, the climate of opinion has been against ad lib feeding through to bacon weight, but improved breeding now enables many strains of pig to benefit from faster growth rate, compared to restricted feeding, without necessarily meaning the disadvantage of poorer grading. Care in design and space allocation is vital because experience has shown that competition, hopper siting and feed freshness can all combine actually to inhibit full appetite, thence being conducive to failure to achieve the most rapid growth. However, with pigs of appropriate genetic make-up, ad lib feeding can prove to have many advantages, particularly in respect to simplicity, layout and low capital costs of equipment. Reduction of waste is a further attractive feature in favour of ad lib feeders. On a commercial unit with a part slatted house floor, feeding was replaced by single ad lib hoppers. Growth rate improved by 250 g/day and feed conversion efficiency by 0.8. There was no evidence of impaired gradings and it was deduced that the difference was due to wastage and spoilage, as some of the feed was going down the slats. In recent ADAS trials on ad lib feeding a difference in feed conversion efficiency of between 0.23 and 0.49

was recorded in favour of single place compared to conventional group feeders. It was assumed that the difference was due to less wastage from the hopper with reduced trough space. Even an improvement of 0.25 in conversion efficiency is worth £2 per pig at today's feed prices. At 18 pigs to one hopper its cost could be repaid in two or so batches.

Self feed hoppers, with feeding space of not less than 75 mm for each pig in that pen, for group feeders are replenished with feed by equipment similar to that used for volumetric dispensing, as previously discussed. Thus, centreless auger or chain/rope and disc conveyors transport feed to a circuit of ad lib hoppers, filling them in turn down drop tubes, and are supplied from a bulk bin situated outside the building.

Pipeline feeding systems

A new generation of pipeline feeding systems now available which incorporate flow meters or load cells combined to microchip technology offer hitherto unobtainable feed dispensing accuracy. This together with renewed pig producer awareness of the value of incorporating a range of byproduct feeds, both liquid and pulped, is promoting increasing interest in wet feeding systems. But what are the overall advantages and disadvantages of wet feeding?

Advantages of wet feeding

(1) IMPROVED PERFORMANCE
Work carried out by the former Agricultural Research Council (ARC) by Braude and others in the 1960s clearly demonstrated improved performance, with dry fed pigs needing 0.2 kg more feed for each 1 kg liveweight gained compared to wet fed pigs, and an average of 10 days less to slaughter for wet feeding. But more recent trials begin to question the accepted advantage of superior daily liveweight gain for wet feeding, suggesting that wet fed pigs can even take 2–5 days longer to reach slaughter weight against dry fed pigs. However as less feed is wasted, feed conversion is still in favour of wet feeding.

(2) LESS WASTE AND DUST
A properly designed and maintained pipeline system should certainly ensure that minimal food is wasted compared to dry feed systems. Furthermore, dust is curtailed – another form of waste – and aggravation to lung complaints, in both stock and man, reduced.

(3) INDICATIONS FOR USE OF HOME MILL AND MIX
Evidence shows that for dry feeding, pellets give better performance than meal. Thus, the implication is that the home mill and mixer who feeds dry meal, even of nutritionally identical specification to a compound pellet,

will experience an impairment in pig performance unless he accepts the inevitably high cost of pelleting on the farm. Wet feeding of meal removes this problem. Further, where pig feeding is the main enterprise, the need for a separate mixer to prepare milled rations can be avoided as all the ingredients can be added directly to a batch type liquid feed mixer.

(4) CONTROLLED ACCURACY
Current pipeline feeding systems offer the facility for very accurate control of rationing. More sophisticated systems can automatically control the amount of individual ingredients, wet or dry, added to the mixer and then feed each pen automatically. A number of feeding refinements can be built in at relatively modest cost, exploiting the full extent of computer control and further increasing the appeal of liquid feeding.

(5) CONVENIENCE AND FLEXIBILITY
Wet feeding offers particular flexibility in raw material use which no other system can provide. Thus, a wide variety of granular, liquid and pulp feeds can be incorporated, such as whey, skimmed milk, potato and brewery waste, with considerable potential for opportunity buying and reduction of feed costs. In the future, the range of such materials may well extend to include root crop pulp, grass juice and other materials available on world markets.

Handling feed in liquid form, despite its bulk, does enhance convenience and flexibility in handling, and pipelines can often readily accommodate the limitations of installation in old buildings. Extensions are relatively simple and extra long pipelines can be serviced by introducing booster pumps along the pipeline. Pipeline feeding, unlike dry floor feeding, is highly suitable for the popular housing system using fully slatted floors. Electrical installation, for major power use, is relatively simplified, as motors involved are centralised at the mixer. At higher dilution rates the need for separate drinkers could be dispensed with, but current Welfare Codes should not be ignored.

Disadvantages

(1) CAPITAL COST
Overall capital cost is significantly higher than for on-the-floor feeding of pellets (typically £10–15 per pig place), ameliorated to some extent for home mixers by the saving in the high cost of pelleting.

(2) CARCASS GRADING
Gradings may be poorer, especially where the daily intake of nutrients has been overdone, through miscalculation. The overall effect is often not economically significant if FCR is satisfactory. The killing out percentage may be reduced as a result of increased gut fill.

(3) NON-SUITABILITY FOR YOUNGER PIGS

Younger pigs under 35 kg are not well suited unless a more concentrated mix is used, as at conventional dilution rates (for example, 1:2.7 by weight) younger pigs would not receive their optimum fill of nutrients before appetite is satisfied. A mix of 1:2.4 is preferable for smaller pigs and this may be difficult to pump round the circuit reliably.

(4) RELIABILITY

Blockages may occur in certain circumstances adversely affecting the overall reliability of a fully automated system. However, with care in design and proper operational practice blockages can largely be avoided, as discussed later.

Unless adequate steps are taken, pipelines can freeze up in frosty weather. If necessary, pipes should be appropriately insulated and/or electrical chase heating tape wrapped round vulnerable sections of pipe.

Fermentation, producing gassing, may result in flooding of feed through automatic valves between feeding times in hot weather unless the system is adequately and regularly flushed through. A very dilute solution of formalin can also cure the problem and cause no harm to the pigs. A 40% solution of formalin diluted 1:700 should be sufficient.

Problems may also be encountered in the storage of food byproduct ingredients. Thus, slurries or suspensions, such as cereal starches, liquid potato and liquid yeast may be prone to settlement and compressed air or pumped recirculation agitation may be necessary in storage.

(5) STRESS

Pipeline feeding can only be carried out on a pen-by-pen basis – hence pigs inevitably get excited once they hear the system start up. A system developed a few years ago by the former National Institute of Agricultural Engineering (NIAE), now AFRC Engineering, Silsoe, provided temporary hopper storage for wet delivered feed with a special unloading plug to ensure effective purging of the hopper and remixing of settled out feed in the trough.

Mixing systems

There are basically two main systems of pipeline mixing and feeding: batch and continuous.

(1) BATCH PIPELINE FEEDING

This is by far the most common option. Here a predetermined quantity of feed is first mixed, then distributed and fed to pigs, either on the basis of one or more large batches, a system which can easily be automated, or on the basis of an automated repeat batch system using a smaller capacity mixer. Though the delays between each batch will extend feeding time,

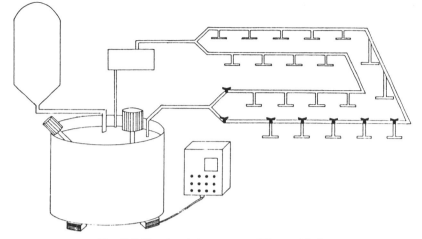

Fig. 3.3 Batch mixing system (*Hampshire*).

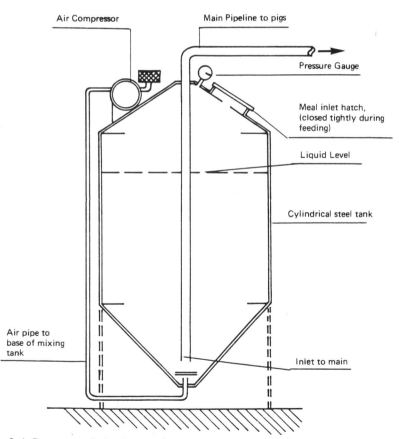

Fig. 3.4 Compressed air pipeline feeding system. (*Reproduced by kind permission of Farm Electric Centre (Electricity Council*).)

this method enhances flexibility in the number of rations that can be fed.

Raw materials can be measured volumetrically, but it is usual for steel or GRP mixer tanks to be mounted on load cells providing a visual indication or, more commonly, automatic control of the filling cycle. In many cases too, the load cell provides signals to control feed dispensing. Mixer tank capacities are typically in the range 1800–9000 litres.

While meal is added to liquid in the tank, mixing and agitation of the tank contents can be achieved pneumatically, but most usually by means of a motor driven paddle. Batch mixing takes 5–15 minutes before the mix is circulated round the distribution main.

Pneumatic batch mixers, though not commonly found today, use an air compressor pressurising a sealed mixing tank with air at about 0.8–1 bar. Feed is then displaced through a 50 mm diameter, single delivery pipe. For mixing, after liquid has been delivered through a top hatch door meal is slowly added while air is bubbled through the liquid to ensure proper dispersion.

Significant advantages of pneumatic mixing and delivery are that pump wear and tear is avoided as no moving parts come into contact with the potentially corrosive liquid, no separate motor is required for mixing and a ring main delivery system may be unnecessary. Problems may be experienced, though, in adequately purging the line of feed after feeding.

Whatever form of batch mixer is in use, it is imperative to remember that mixer tanks should not be entered unless absolutely necessary due to the risks of asphyxiation from carbon dioxide gas. The vessel should be emptied and ventilated, and help should be standing by with rope attached to the person entering.

(2) CONTINUOUS MIXERS

Continuous mixing and distribution is somewhat cheaper to install but gives less flexibility in opportunity to use different feeds. Such systems are unsuitable for the handling of slurries (such as cereal starches or liquid potato) or some solids like biscuit meal unless pre-mixed. Various types of continuous mixer are in use, all working on the same principle: measured quantities of feed and liquid are metered into a relatively small mixing tank of 100–150 litres or less, then instantly mixed and distributed. As feed is removed a level probe or load cell is employed to control the addition of further ingredients in a continuous process. The upper limit for the number of pigs capable of being fed is a function of the rate at which ingredients can be added by auger from the feed hopper and by pump from an adjacent header tank.

Versions differ in the means of control of quantities added and method of mixing. For most, ingredients are measured in on a volumetric basis in pulses (for example, A & H and Hydrop) and for instance, a meter in the water line sends a pulse to the control panel to start and stop the meal discharge auger as each desired quantity is mixed in the correct proportion. Another system uses load cells to weigh in constituents. For mixing,

Plate 3.5 Continuous mixer. (*Hyde Hall Ltd.*)

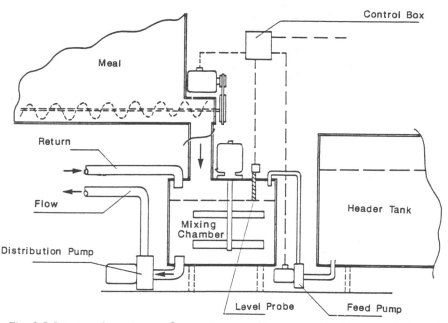

Fig. 3.5 Layout of continuous flow wet mixer. (*Reproduced by kind permission of Farm Electric Centre (Electricity Council).*)

some systems use a large multivane paddle while another uses the turbulence effect of adding water and meal together in an appropriately sized small mixing hopper which is mounted immediately over the distribution pump suction point (A & H). The NIAE vortex mixer operates in a similar way. Developed several years ago, this mixer is not produced commercially in the UK although limited production has taken place in the USA and New Zealand.

Pipeline distribution

Most systems are designed to cope with meal/water mixes of 1:2 to 1:3.5 by weight, although around 1:2.5–1:2.8 is the most common dilution rate adopted. Except for systems where compresed air is used to distribute feed, the norm is to utilise either single stage centrifugal pumps or eccentric helical rotor pumps. The latter are capable of dealing with longer circuits or more viscous feed. Self priming centrifugal pumps are most commonly used with a bronze or stainless steel rotor, which will usually need replacement every 1–2 years. Typical pump output may be around 500 l/min at 18–20 m head, with a pump power range of 2.2–11.2 kW. The pump may be either close coupled to the motor and mounted on top or to the side of the mixer tank or fitted on the end of a long drive shaft and located in the bottom of the tank. This facilitates pump suction and

Plate 3.6 Feed pump/agitator (*Hampshire*).

it is common to fit another agitator impeller above the pump housing to provide additional agitation at the suction point.

The speed of filling of the mixer can be improved by the use of multi-port valves, porting the pump to draw water or other liquid material from storage tanks into the mixer vessel.

Steel piping has now been almost entirely superseded by rigid uPVC of 50–63 mm outside diameter for distribution mains and 38 or 45 mm diameter for branches to individual pens. uPVC piping is corrosion free and cheaper than steel, but it does require more frequent support (for example, 450–600 mm centres) when used overhead, which offsets some of the savings. Steel, for overhead use, needs supporting every 3 m. Vulnerable sections of piping may need to be protected against frost by lagging and/or incorporating chase heating tape at around 15 W/m. Alternatively, the pipe may simply be drained off in frosty weather.

In terms of distribution piping layout, although some feeding systems will accommodate a single dead-end mains supply pipe, most use a flow and return layout where several separate circuits can be arranged for

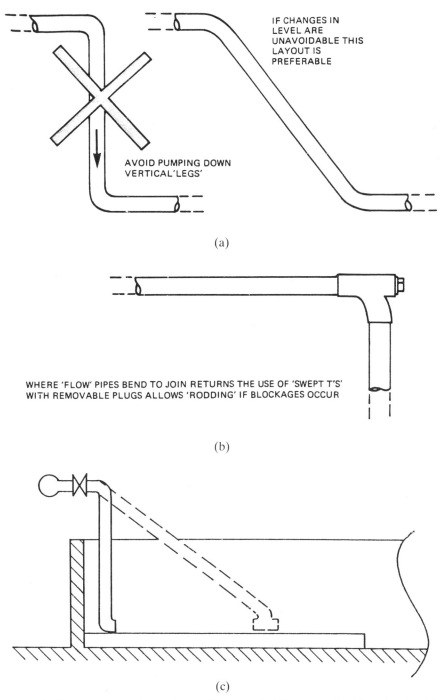

(a)

(b)

(c)

Fig. 3.6 Design aspects to minimise blockages/settling out. (a) Slow bend. (b) Swept tee with rodding plug. (c) Location of valve outlets to minimise settling out.

different houses or mixers. Each separate branch off the main is controlled by a distribution valve. On automatic dispensing systems a valve is also provided on the return line which is closed when feed is being distributed to individual pens. In one system (Funki) a throttle valve is incorporated into the distribution flow near the pump and controlled by the central control computer. This allows for pressure compensation when feeding from the nearest trough to the furthest and improves dispensing accuracy so helping to prevent splashing and spillage.

Potential blockages can be minimised by care in design detail and proper operational procedures. The most important design aspects are:

- The pump should be of adequate capacity for the length and diameter of pipe involved.
- Pipe diameter should be adequate for the flow involved (for example, 50 mm diameter for a typical ring main should cause few problems).
- Sharp changes of direction must be avoided (by using, for example, slow bends).
- There should be no drop from the main to each feeding valve, which would allow settling out.
- Appropriate falls should be incorporated to assist drainage and minimise pockets where settling out may occur.

Operational factors to reduce the likelihood of blockage include the following:

- Particle size should not be too large (for example, barley milled to grist of not more than 3 mm).
- Excessively thick mixes should be avoided.
- Correct filling procedures should be adhered to, that is filling with liquid first before adding meal.
- Pipelines should be flushed regularly (preferably after each feed). After flushing with water the flushings may be returned to the mixer tank. However, where these may contaminate a subsequent different diet, a separate rinse tank may be provided adjacent to the mixer tank.

Feed dispensing

MANUAL

The choice is either to have a single fixed valve for each pen or a shared outlet point with plug-in hose dispenser. Fixed valve outlets use either tapered plug cocks or, better, seat stop ball valves, the latter being less prone to leakage. Brass valves are giving way to corrosion free uPVC seat stop ball valves, often with low friction PTFE seats.

Rationing is achieved by eye on a timed basis or, better, by use of a flow meter. Some systems incorporate a large digital display or displays mounted prominently in the piggery to assist the stockperson in rationing.

The display is fed from signals from load cells beneath the mixer or from a flow meter in the flow main. Such displays can be self zeroing when flow ceases. The end of the return pipeline is normally valved off during feeding, after initial circulation.

Flow meters are of two types: electromagnetic with no moving parts or rotary. Some of these are capable of being retrofitted to existing layouts as long as a centrifugal pump is used, but caution is needed with long or tortuous circuits where trapped air pockets can cause problems with flow meter 'run on'. A portable plug-in type flow meter is also available from one manufacturer with a 'petrol pump' type dispenser.

The benefit of dispensing by weight lies in avoiding complex calculations in reconciling weights of ingredients mixed with quantity of mix feed.

AUTOMATIC DISPENSERS

Here each pen is served by an automatic valve normally kept closed by a compressed air system and opened electrically by solenoid air valves. An airline and wiring 'harness' is thus clipped alongside the pipeline main.

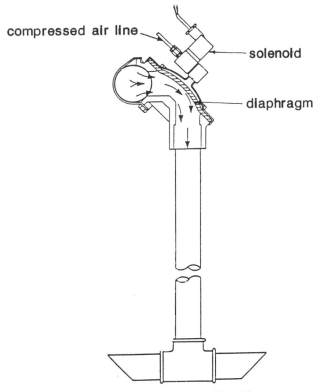

Fig. 3.7 Diaphragm valve (*Funki*).

For feeding to take place, once the return line valve has been automatically closed signals received from load cells, or a flow meter, enable the controller to open each valve in turn to release feed.

With some systems the main is kept full between feeds with the compressor running intermittently to top up an air reservoir, keeping valves closed. On another computer controlled system, the option is provided to give automatic flushing with clean water at the end of a feeding cycle.

Valves are normally either of two types: diaphragm or 'squeeze' type, although piston and automated ball types have also been used in the past. Whatever valve is used, it is obvious that maintenance needs should be few and that repair or replacement can be carried out quickly. Whilst the pinch valve is a total replacement job with quick hand union change, the rubber in a diaphragm valve is changed by removing four bolts.

For emergencies, it is important that the solenoid valve associated with each valve can be overridden manually.

Control

Three levels of control are used: manually assisted, semi automatic and computer controlled.

Manually assisted systems often incorporate load cells to facilitate manual control of feed preparation, with a digital display at the control panel to give a read out of feed added, and a further digital display in the piggery to assist manual dispensing.

Plate 3.7 Pipeline feed computer controller (*Hampshire*).

Semi automatic systems provide the facility for either automatic feeding and manual feed preparation or, more usually, the opposite. Thus one system, for instance, allows for automated preparation of up to four diets (or pig groups) with up to twelve different ingredients in any combination up to three times during the day. Distribution is under manual control with a two-digit display with a claimed ±1 kg accuracy for both preparation and distribution.

Computer control, as discussed earlier, not only gives full automation of preparation and feeding, but can also provide for highly sophisticated refinement. One such system, in addition to providing a similar specification to the semi automatic system described above for feed preparation, can automatically control, within very acceptable limits of accuracy, the operation of up to 256 valves through various circuits and with automatic run-on corrections to enhance accuracy. Other facilities include totalling of the amount of feed needed to establish the quantity to be mixed and automatic adjustment of feed for each pen as pig weight increases. In-built feed formulation and costing programmes are also available with the ability to total amounts and costs of feed input per pen over a specific period.

Mix at trough systems

Mix at trough systems have been popular in North America for some time and are now becoming progressively more popular in Europe. In principle, the system attempts to balance the advantages of wet feeding with the lower capital costs of conveying and dispensing feed in dry form with the opportunity for pigs to wet and possibly mix the feed themselves at the trough. The so-called 'soak system' is an unusual variant to this general principle in that once dry feed has been dispensed it is wetted by spray bar and the mixture left for some time to soak before being presented to pigs.

Recent North American and British trials have shown significant improvements in FCR and growth rate with this technique compared to conventional hopper systems of feeding. Apart from the soak system there are four main systems commonly used in the United Kingdom:

(1) AD LIB, PIG OPERATED, BULK HOPPER FEEDERS WITH DRINKING NIPPLES (FOR EXAMPLE, AQUAMIX AND PYRAMIX)

These are relatively simple and cheap, comprising a bulk hopper with trough beneath and built-in nipple drinker(s). Feed is delivered by gravity when the pig roots about and pushes on a displacement bar which, when nudged sideways, delivers a small parcel of food. Alternatively, the pig chews at meal through a delivery slot running the length of the trough. The idea is that the pig will alternately drink and eat dry meal at the same

Plate 3.8 Pig operated point feeder (*Funki RA-DOS*).

place. Water escaping from nipple drinkers enables the pig to mix its own cocktail of gruel in the trough.

The main problem with these types of feeder is that food can get stale in the bulk hopper.

(2) RESTRICTED AD LIB, PIG OPERATED, POINT FEEDER WITH TROUGH AND DRINKING NIPPLE WITH AUTO CONTROL (FUNKI RA-DOS AND COLLINSON QUANTUM)

These types of feeder are rapidly gaining in popularity. One version comprises a series of small hoppers delivering feed down a narrow transparent plastic tube to a pig operated 'tongue' or flap dispenser mounted above a small trough also incorporating a nipple drinker.

The small hoppers are kept replenished by a cable and disc conveyor from a master feed hopper into which feed is automatically weighed from an outside bulk food bin. Feed is delivered when the pig lifts the flap, the degree of opening being adjustable according to feed type. As in the previous system, by operating the nipple drinkers the pig mixes feed to its own requirements. The control panel can be set with the required number of pauses in availability of feed in any one period of 24 hours, as well as the maximum amount of feed to be distributed. The automatic weigher also enables a tally to be kept of the amount of feed delivered.

In addition to providing closer control over quantities fed, the system also integrates the benefits of a convenient form of feed distribution with little risk of feed going stale.

In another version, by pushing a paddle wheel across the back of the trough, pigs drive a small dispensing auger which delivers a small amount of feed from the hopper above into the two sides of the trough. Water is available to the pig by means of a nose operated spray type drinker situated beneath the dispenser at the back of the trough. Feed can be restricted using a simple graduated brake mechanism acting on the shaft of the paddle wheel dispenser. If pressure on the shaft is increased, the paddle wheel becomes increasingly difficult to rotate and as a result the pig will rotate the paddle wheel fewer times per visit.

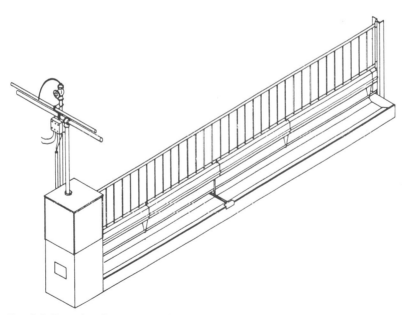

Fig. 3.8 Restricted cross trough feeder with spray bar (*Funki Flexi-DOS*).

(3) RESTRICTED CROSS TROUGH FEEDER WITH SPRAY BAR AND AUTO CONTROL (FUNKI FLEXI-DOS)

This system can accommodate feeding in a trough up to 6 m long. It comprises a longitudinally divided trough with access both sides shared between adjacent pigs and served by an auger which distributes feed at around 50 g per 30 cm trough length through a longitudinal slot along the length of the trough beneath. Feed is then wetted by a water sprinkler bar. The control bar automatically controls water and feed quantities, the amount fed and duration of pauses between feeding sessions. The system can permit up to nine feeds in a day.

(4) RESTRICTED AD LIB PIG OPERATED PAN FEEDER WITH DRINKERS AND AUTO CONTROL (CHORE TIME TURBOMAT)

This is a system very popular in mainland Europe and also increasing in popularity in the UK. It consists of circular pan feeders divided into ten feeding places, each pair sharing a nipple drinker and each pan feeder serving a pen of around twenty pigs. Feed is conveyed overhead by a circuit of 55 mm or 75 mm diameter flexible auger from an outside feed storage bin. Feed is supplied to each feeder down a drop tube incorporating feed level adjustment. The pan incorporates a simple rotor with which the pigs have to root for the feed to be delivered into the pan. Pigs tend to alternate shorts bouts of feeding with drinking and spilt water is collected in the pan. The controller is normally set to make feed and water available over three separate periods from 06.00 hours to 20.00 hours for older pigs and more frequently for younger pigs. Normally, availability periods are set at 1.5 hours for feed and 2 hours for water. Restriction of water, although not strictly complying with Welfare Codes, does, it is claimed, reduce volumes of slurry collected. Other particular features of pan feeding are its ready application to fully slatted floor systems and significant economy in use of floor space by pan feeders compared to feed troughs. Regular periods of free access followed by rest periods are said to minimise stress. Although relatively expensive compared to simple mix at trough systems, recent pig producer experience has demonstrated encouraging performances from this system.

Liquid baby pig feeders

Automatic supplementary liquid feeding of baby pigs is a practice popular in the USA and now introduced into the UK which is proving to be beneficial in boosting weaning weights. Potential exists, too, in saving runt piglets and enhancing numbers of pigs weaned per litter. It is a system designed to supplement the feeding of creep feed pellets, but some producers have actually abandoned solid feeding of the litter in favour of liquid. One system comprises a 127 litre glass fibre tank situated in a cool area outside the farrowing room, from which freshly made up milk sub-

Fig. 3.9 Mix at trough pan feeding system (*Chore-Time Turbomat*).

Plate 3.9 Mix at trough pan feeder (*Chore-Time Turbomat*).

stitute is pumped through a 22 m PVC pipe to supply a series of 'farrowing cups', one or two per farrowing pen located in the creep area. Each cup, fitted with an on/off valve, is specifically designed to reduce milk wastage and keep piglets clean. It has been found that each litter consumes on average about 1–1.5 litres per day and about 2.5 kg of milk powder before weaning. A mini version is also available comprising cup and fittings designed to be attached to a bucket above the pen.

Another system consists of an automatic liquid feed mixer and dispenser built into a pan feeder. Every hour a 'meal' is prepared when milk substitute powder is augered to the blender where mixing with water takes place. A time clock regulates the number and length of feedings and controls the rinsing of the blender with water after each feeding. A fresh supply of milk powder is added to the feeder twice a day.

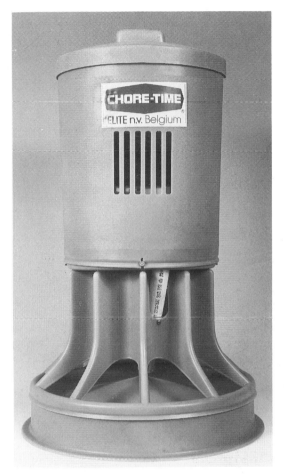

Plate 3.10 Baby pig feeder (*Chore-Time*).

Mechanised sow feeding

Table 3.4 presents the main current choice of mechanical feeding systems for dry sows.

Individual sow feeders

The necessity for individual feeding of dry sows is, of course, completely accepted in respect to protection from bullying and providing means to control feed intake. Traditionally, dry sows have been fed individually by hand in sow feeder crates. It is a relatively easy step to mechanise and possibly automate feeding in such stalls using either pipeline or dry feed point feeders as used for fattening pigs.

For dry feeding a range of scaled down point feeders are available for

Table 3.4   Mechanised dry sow feeding systems.

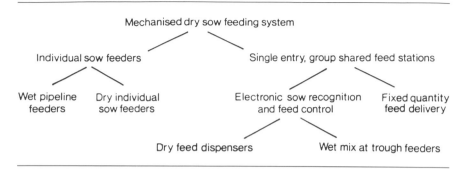

Plate 3.11 Feeding system for sows (*Chore-Time*).

sows, virtually all controlled volumetrically. Thus, an overhead centreless auger or chain/rope and disc conveyor supplies a row of small hoppers typically fitted with a telescopic or regulated slide type ration adjustment for each feeder. In one system dispensing is effected by a pull on a wire

Plate 3.12 Flat rate electronic sow feeder (*Pig Rigs Rotafeed*).

rope connected to each feeder or by lifting a discharge tube allowing feed to fall down the drop tube to the manger. The cable may be controlled by either hand winch or a time controlled motorised winch.

In another system, delivery is effected by rotation of a common long rod around which a cord is wound to lift plastic bungs, one for each feeder. The rod is rotated by either a hand operated winder and gearbox or a time clock controlled motor and gear box.

Such feeders may also be used for lactating sows.

Single entry, group shared feed stations

(1) FIXED QUANTITY FEED DELIVERY TYPE

This low cost feeding system relies on a two pen layout. Sows to be fed are moved into pen 1 from where they each move into a single entry feed station when they are locked in. Each sow in the group receives the same quantity of feed. After feeding, each sow moves through the front exit gate into pen 2. Sows needing a different feed allowance are put into a second group where the feed station is adjusted to a different daily allocation. Flat rate feed level is adjusted by push button on the built-in control box.

(2) ELECTRONIC RECOGNITION AND FEED CONTROL SYSTEMS

Electronic sow feeders, developed from out of parlour dairy cattle feeders, are no longer a novelty. Indeed, there has been a comparatively rapid and widespread adoption of this system of controlling sow feeding, often associated in the UK with more flexible loose housing in straw yards. Heralded as a breakthrough by animal welfare organisations, electronic sow feeding, in the right hands, permits considerably more sow freedom and potential reduction of stress but also provides opportunity to feed individual rations to group housed sows at a cost per sow place equivalent to conventional sow housing.

General principles of electronic sow feeders

Sows are fitted with sealed electronic responders or transponders normally incorporated in a neck collar or ear tag. This enables each sow to gain access to a feed stall where her electronic number code is 'read', signalling to the controlling computer to dispense an appropriate amount of feed at predetermined intervals. Feed may be dispensed in dry form or wetted by a water spray at the manger. Experience has proved that accuracy of

Plate 3.13 Electronic sow feeder (*Collinson*).

dispensing is very acceptable and that somewhere between 40 and 55 sows can share one feed station, but observation suggests that 30 to 40 sows per station create less aggression. The feeders are best suited to large herds, with a minimum of three feeders on the unit.

Housing can be arranged with single feeders serving groups of sows or gilts weaned in batches. This avoids the mixing of sows and the increased levels of aggression associated with it. But for most herds mixing is necessary and here aggression is minimised in a large group as there is more space available for a fleeing sow. Thus, most prefer to have all the sows in one group with several feeders. One Essex farmer has experience of housing 400 sows together, without detriment, in a single large strawed yard.

Advantages of electronic sow feeders

(1) GROUP HOUSING ON STRAWED YARDS
The system enables sows to be group housed on straw in large numbers without detriment to control of individual feed intake with the attendant advantages of:

- Cheaper housing.
- More unrestricted space for sow exercise.
- Sows can choose their own 'environment' within the yard.
- Reduction of sow stress and promotion, in practice, of more docile sow behaviour.
- Meeting the spirit of the Welfare Codes.

(2) CAPITAL INVESTMENT
The combination of strawed yards with electronic feeders produces a system which is often no more expensive than more conventional housing and feeding systems based on stalls or tether systems.

(3) EQUIPMENT UTILISATION AND SPACE NEEDS
Compared to hand operated individual sow feeding crates, electronic sow feeding stations are not only better utilised but require considerably less space, as each station occupies the area of only one crate.

(4) FEEDING CONTROL
Electronic control enables easier feeding of sows according to their condition and state of gestation within relatively fine limits of accuracy, promoting efficient use of food and high standards of sow performance. Furthermore, feed regimes can be instantly adjusted by a predetermined amount for all the sows when, say, there is a marked change in weather conditions.

(5) REDUCED LABOUR INPUT FOR FEEDING
By removing the routine noisy chore of hand feeding the stockperson can devote more time to stockmanship.

(6) ENHANCED MANAGEMENT INFORMATION
As well as directly controlling the dispensing of individual sow feed, a check is automatically made of those sows which have not received their daily quota. Computer control also lends itself as a built-in management information system, maybe obviating the need to purchase a separate stand-alone system and possibly giving the opportunity for substitution of sow cards. Not only can performance records be stored on computer file, but also the system can incorporate automatic 'flagging' of individual sows' oestrus and farrowing dates, for instance. On the appropriate date that dry sows should be drafted to farrowing accommodation they can be automatically sprayed with a colour marker whilst visiting the feed station, so improving identification for the stockperson.

Disadvantages of electronic sow feeders

(1) CAPITAL OUTLAY
Although overall feeder and housing costs compare very favourably to stalls/tethering systems, there are often definite cost benefits deriving from the continuation of use of an existing system which may often be cheaper than capital outlay in a new system. In a volatile period for pig marketing there may be good reason to show caution in further investment.

(2) VULNERABILITY OF FEEDERS
Although the current generation of feeders is more sturdy, springs and catches are still prone to cause some minor reliability and maintenance problems. Experience suggests that the electronics are pretty reliable but enemies such as fluctuating voltages and temperature, damp and dust can be a problem. Agents must provide service within 24 hours since delays are unacceptable. A reliable electrical supply is, of course, fundamental, and a stand-by generator will be a useful adjunct.

(3) SOW TRAINING
A well orchestrated sow and gilt training system is needed with training pens and adequate stockperson time available to ensure that all sows and gilts quickly learn to use feed stations. This can involve a significant time input, particularly with timid sows/gilts. Even then a few sows refuse to accept the system and will need to be accommodated separately – or sold.

(4) STOCKPERSON'S OBSERVATIONS
With group housing it is less easy for the stockperson to recognise and adequately observe the behaviour of individual sows. Thus, greater responsi-

bility is thrust upon the stockperson who, ideally, should walk through the yard daily, get each sow on to her feet, give her a close inspection and make handwritten notes on her record card. Hand held electronic sow recognition units are now becoming available with LED display of sow number and which might, in the future, also incorporate access via a keypad to sow records and perhaps allow data input, for later downloading to the master controller.

Once the feed drop mechanism operates, the computer registers that feed has been dispensed. However, if the hopper is empty the system will not recognise that the sow has not received her feed. Blockages are not always easy to spot, thus good stockperson vigilance is vital for successful operation.

(5) ENCOURAGEMENT OF UNDESIRABLE SOW BEHAVIOUR

With some feeders, sows may be reluctant to leave the feed station with the prospect of more feed becoming available. Although research suggests that sow loitering may be on average three minutes post-feeding it could be up to half an hour and in some early versions some sows were even prone to sleep in the feed station. Current well designed versions overcome this problem by locking the sow in for a period appropriate for her to clean up her feed, following which the exist gate is released and shortly afterwards the entry gate is unlocked so that a following sow may drive her out.

Experience shows that in yarded sows vulva biting remains a problem, encouraged when sows congregate eagerly around the feeder waiting their turn. Rear entry feeders, with front or side exit, as distinct from earlier rear entry and rear exit feeders, can do much to control the problem as well as care in the design of entry gate to give adequate rear end protection to the sow.

Feeder equipment

Equipment is similar to out of parlour dairy feed dispensers, comprising:

- Stall.
- Feed dispenser.
- Ancillaries.
- Control system.

STALL

Experience has shown that the stall needs to be one hundred per cent sow proof and extremely robust. Rear entry and rear exit stalls tend to encourage bad sow behaviour so that most current versions are based on 'walk through' rear entry and front or side exit layouts. Entry and exit gates are either of the up-and-over type or side hung, in single or double format, often incorporating a vertical roller bar or series of rollers to ease sow access and make the door more pig proof. Doors are often linked with a bar so that the opening exit door also opens the entry door, allowing the next sow to enter. It is usual for doors to be designed to inhibit two sows from trying to enter together and for the exit race to be fitted with one way escape gates, thus easing sow exiting.

Gate control type splits stalls into two basic groups:

- Sow operated.
- Electronically/pneumatically operated.

With sow operated gates, as the sow enters the stall she causes the rear door to close and lock behind her, effected in one version by her 'snouting up' a counterbalanced up-and-over door arrangement. The sow remains in the stall for as long as she needs to feed then, on leaving, she will operate the exit door and release the lock mechanism, opening the entry gate for the next sow. This system most readily accommodates individual variation in eating speed habits, but there is a danger that sows may lie inside for long periods blocking access to others sows, with the possibility that on a widespread scale cumulative delay will adversely affect proper opportunity for feeding. Provision of an anti lying bar reduces this likelihood.

With electronically controlled versions, gates are opened and closed by pneumatic rams supplied by a small air compressor. The entry door is automatically opened, allowing a sow to enter and to be locked in. If she is entitled to feed the gates remain locked and meal or pellets are dispensed. At a predetermined time, once she has completed feeding, the exit gate unlocks allowing her to vacate the stall and another sow to enter. Where a sow is not entitled to feed, gates are opened and a following sow will coerce the first to leave. Computer controlled gates need very careful adjustment to allow for individual variation in feed consumption.

FEED DISPENSER

The feed dispenser comprises a feed hopper supplying an electrically driven auger metering device with adjustable output. In one version the auger fills a small plastic box which is intermittently inverted, releasing feed to the feed trough, while in another the auger supplies feed directly to the trough and rotates at a rate of one revolution every 15 or 20 seconds until the sow leaves the feed station or has received her whole particular entitlement at one sitting.

It is important that the drop interval is set so that each sow has time to consume most or all of her feed before the next portion is dispensed. This is especially critical with electronically controlled doors to ensure that all feed is consumed before the exit door is opened. Any feed left unconsumed will encourage premature feed station revisiting by sows without entitlement. Calibration must be carried out regularly due to different feed densities between batches. Actual calibration procedures vary considerably, some being easier than others. One version, however, incorporates an 'automatic' calibration routine where the stockperson only has to weigh ten drops of feed into a special bag and enter the amount in the computer. A simple and quick routine will mean that calibration is more likely to get carried out.

Most feed stations use a fixed manger constructed of steel, concrete or corrosion resistant moulded polymer concrete. However, one Dutch feeder uses a novel revolving feed trough which is moved, under computer control, to open up access to either the normal exit race or to an isolation area.

Except in isolated locations, feed dispenser hoppers are normally designed to be automatically replenished by centreless auger or chain and disc conveyors from a nearby bulk storage bin.

ANCILLARIES

Most feeders are available with optional facilities such as wet feeding, auto marking and auto sow drafting. Wet feeding is achieved using either

Plate 3.14 Auto spray marking system (*Collinson*).

a water injector device, which allows water to be added to meal in predetermined ratios, or a wet feed dispenser with diaphragm valve to dispense predetermined quantities of wet mixed feed.

Most feeders are now available with facilities for automatic spray marking of appropriate sows – for instance those sows computed ready for removing to farrowing accommodation. This is simply achieved using a reservoir of dye and a propulsion system from the built-in compressed air system.

Some feeders are also available with double exit races with means of automatic drafting of appropriate sows into an adjacent area, where important tasks such as pregnancy testing can be carried out. Further options include a trough flap to discourage sows not entitled to food and a device for restricting crate width, to prevent gilts from attempting to turn round.

CONTROL SYSTEM

The control system consists of three main components: an encapsulated responder or transponder worn by the sow in a neck collar or ear tag, a radio wave transmitting and receiving system and the control computer.

Many systems are based on an encapsulated active responder, with long life lithium battery, worn on the neckband or fitted into a plastic ear tag, together with response transmitter. Normally in a quiescent state, when the tag comes within range (normally within 600 mm) of the radio frequency field produced by an interrogating antenna, it is activated and transmits its data back to the antenna. As electrical power use is small and intermittent, battery life can be expected to be several years. The animal's identification number is then sent to the computer to ascertain whether the sow is entitled to feed and, if so, an appropriate amount is dispensed.

A number of feeders, however, use a passive transponder system similar to that used for identification of dairy cows, with an encapsulated transponder normally fitted to a neck collar. It is necessary with this system for the sow to come into closer proximity to the transmitter and receiver system. This is often achieved by arranging a ring aerial built around the opening to the feed trough.

Much debate has been devoted to the various virtues of sow collars and ear tags. Originally collars were the norm but as electronics have become miniaturized a number of systems now use ear tags which, although less easy to fit compared to collars, are less prone to being lost and avoid the likelihood of neck lesions. Ear tags are not completely loss-proof, however, and while a hand held location device to search for them in straw is feasible, the audio range of the electronic tag is somewhat limited.

However, with still further miniaturization, responder implants have become a reality. Normally surgically inserted on the neck behind the ear, the implant gives the advantage that it cannot be torn out or lost, but its size does limit its power and its range is consequently limited to around 150–200 mm. In practice, by careful design the sow can be made to present her head sufficiently near to the interrogating antenna to make the system successful.

The latest Belgian system, however, avoids the need for sows to wear any form of transmitter. It uses a computer controlled video camera mounted on the feed station to identify animals. Yes, there are sufficient differences between all animals in the herd to make the system work!

Stockperson communication with the control computer, in basic format, is via a keypad and LCD display. However, more sophisticated versions, using video display unit, disc drive and printer, give faster and more comprehensive data access and a greater degree of flexibility. Most software incorporates 'management by exception' type programs to print out daily which sows have not fed, to alert the stockperson to sows which have either lost their collars/ear tags or are 'off colour'. Though most programs arrange that a sow receives all her feed at one visit, some also allow the opportunity to divide feed entitlement into small portions giving a 'little and often' regime if need be. Each computer may have the capacity, typically, to serve 6–12 feeders and up to 400 sows, but some are even capable of handling exceptionally up to 64 feed stations and 2000 animals. Adjustment of feed regimes is carried out by manual input but several computer feeding programs now include a so-called 'sow calendar' so that feeding can be automatically adjusted according to stage in pregnancy. These can offer action lists for returns to service, pregnancy testing or vaccinations due on an individual sow and herd summary basis. In fact, many programs permit comprehensive processing and storage of most pig herd performance data. Automatic sow drafting and/or spray marking can readily be incorporated into the control program. The level of sophistication and degree to which the feed computer is integrated with other management functions is a matter of choice for the individual producer.

Siting of feed stations

Feed stations are centres of particular activity with competition for entry where aggressive behaviour can be much reduced by attention to detail. Thus, stations are best sited away from the main lying area and not on main routes say to water points, with plenty of space around them. Exit arrangements should, ideally, divert fed sows away from the feed station.

Electronic sow feeders have certainly gained rapid popularity in Europe since they were first introduced during the early 1980s. They should definitely not be considered as a labour saving device, but with good stockmanship they certainly do offer the prospect of keeping a tight rein on sow feeding, with computer sow recording and 'flagging' a valuable built-in bonus to aid management control. But perhaps equally important for many producers is the considerably enhanced freedom, not only in housing choice but particularly in terms of animal welfare. They are here to stay. Current models are certainly more robust, reliable and 'pig friendly' than formerly and with continued development in electronics we are bound to see yet more sophisticated control features come along, including the possibility of auto sow weighing and semi auto condition scoring.

References

ADAS (1987) *Electronic Feeding for Dry Sows*. P3085.

Bal, A. (1986) 'Automatic sow feeding is gaining ground'. *Pigs*. Misset. July 1986.

Brade, M. (1988) 'Electronic sow feeders are here to stay.' *Pigs*. Misset. January/February 88, 20–24.

Gadd, J. (1986) 'Electronic sow feeding has the future'. *Pigs*. Misset. May 1986, 12–15.

Gadd, J. (1987) 'Is mix at trough pig feeding really on?' *The Home Mixer* 2(2), 32–34.

Gadd, J. (1988) 'Mix at trough feeding, a quiet revolution.' *Pigs*. Misset. January/February 88, 26–27.

Gadd, J. (Sept 1988) 'Responding to Responders.' *Pig Farming*, 24–26.

Gray, J. (1987) 'Computerised sow feeding on a large herd.' 'Pork and Chips' conference paper. New Buckenham Pig Club and Suffolk Coastal Pig Discussion Group.

Lambert, I. (1987) 'Feeding systems for pigs and poultry.' *Farm Buildings Association Winter Conference Report*, 23–27.

Legters, W. (1987) 'Mechanical liquid feeding.' *Pigs*. Misset. May/June 1987.

Patterson, L.M. (1986) 'Electronic sow feeders.' *Pig Pointers* **12**, 2–6.

4   Mechanised fodder feeding

Introduction

Fodder feeding is a key operation with cattle enterprises throughout the winter months. Forage quantity, and quantities consumed, can have a significant impact on animal performance and unit profitability. Economic pressures have encouraged a growing emphasis on expecting a higher percentage of the animal's metabolisable energy dietary needs to be contributed by bulk feeds, which necessitates maximum intakes of palatable, high quality material. Benefits of simplified feeding systems are now proven and include flat rate concentrate feeding of dairy cows, sometimes achieved by complete diet feeding, combining the feeding roughage and other feeds in one operation. These developments, plus the gradual swing away from self feeding of silage and the wider practice of opportunity purchasing of by-product bulk foods have promoted a considerable expansion in the range of fodder feeding equipment now available.

The actual fodder feeding system adopted inevitably depends on a considerable number of inter-related factors including the number and type of stock, size and layout of the farm including buildings, overall feeding policy deemed appropriate for that enterprise, degree of labour input/automation required and not least investment returns. Any mechanised feeding system must be totally reliable and straightforward to ensure that fodder is supplied so that intake is not restricted below the planned level and, at the same time, physical wastage, deterioration, and contamination of feed are minimised. For the workforce the system adopted should remove tedium, minimise time input spent on feeding operations and ideally facilitate feed rationing and promote accurate measurement of quantities of ingredients. Not least, the system must be cost effective for the size of enterprise; thus mixer wagon systems are normally considered inappropriate for herd sizes below around 150 cattle.

For mechanical feeding, fodder can be conveniently divided into two categories of materials:

- Fibrous, relatively non free flowing bulky materials, including silage, hay and straw.
- Other, relatively free flowing bulky feeds, such as roots, crop by-

products including brewers' grains, sugar beet pulp, stock feed potatoes, and so on.

Silage is the most popular example of the first category of materials. Over the past decade or so there has been a marked trend towards conserving fresh grass as silage and away from hay. This has been encouraged not only by the difficulty in a typical summer in our temperate climate to make good quality hay, but also by the fact that complete mechanical handling, and in particular feeding, is easier to achieve with silage.

Quantities of silage to be fed over, say, a typical 200 day feeding season will depend on the age, weight and class of stock and amount of other feeds incorporated in the diet. Where cattle rations are fundamentally silage based one can expect consumption in the range 4–5 kg dry matter (DM)/day (0.8–1 tonne DM/winter) for 200 kg steers and up to perhaps 12 kg DM/day (2.4 tonne DM/winter) for a 600 kg dairy cow.

Dry matter is an important determinant of intake. Thus, for example, a cow may eat 50 kg of 20% DM grass silage which is the same to her, in dry matter terms, as 25 kg of 40% DM silage, with consequent enhanced capacity to boost DM intake. In terms of handling and storage space requirements, each tonne of silage DM is equivalent to around 3.33–4 tonnes of wilted silage of 30% and 25% DM respectively and about 5 tonnes of fresh material, with a density of settled conserved material of around 350–500 kg/m^3 depending on whether it is cut by simple flail (the lower figure) and the degree of consolidation.

Silage is conserved and stored in one of three ways. Storage in horizontal silos remains the most common technique. In this way silage can be directly fed from the feed face (self feed), with top layers from high silage clamps thrown down by hand (easy feed) or, more usually now, unloaded and transported to another area for feeding in some type of mechanical feeding system.

Big bale silage, whereby bales are conserved by being wrapped in plastic or sealed in bags, has become a particularly popular practice for smaller dairy herds in recent years. Bales are either self fed when placed in circular feeders, for instance, or unrolled mechanically and fed along a feed barrier.

Tower storage of high DM, high quality silage lends itself to semi or fully automatic feeding. Here material unloaded from the tower is distributed by either mobile forage box or mechanical fixed feeders. Investment costs are high and the system tends to be more mechanically vulnerable, both factors contributing to a marked decline in interest being shown in this storage system today.

Feeding silage from horizontal silos

Self feeding

Self feeding of silage directly from (normally) a walled bunker has been a particularly popular technique for many years. It is most effective where there is adequate access from winter housing with sufficient quantity of palatable material to provide ad lib feeding, and with enough bunker width to allow not less than 200 mm feed width for each cow. Self feeding avoids not only the need for time and investment in costly materials handling but also the possible need for additional fodder feeding arrangements. However, good organisation and management are needed to minimise physical wastage and deterioration. These potential problems, the lack of flexibility to incorporate other feeds easily, plus several other limitations – not least the lack of sufficient feed face width for larger herds – have combined to make mechanised systems increasingly popular.

Benefits of mechanical feeding

Mechanised feeding offers a number of advantages:

- Livestock performance may be improved in comparison to self feed, as stock are better able to eat to appetite rather than have intake restricted.
- There are fewer limitations on bunker size. Thus, silage can be stored to a greater height than for self feed silage, provided the walls have adequate strength to take the higher loads imposed by material and associated machinery. Many unloaders can cope with heights of 3 m or more. In turn, this means that bunker width does not have to be geared to stock numbers. Greater storage height of storage produces benefits in three ways. Self consolidation of material is increased with height which assists to exclude air from the mass, and this enhances bulk density. Furthermore, a squarer profile reduces the surface area to be sealed and that also reduces potential wastage. Less floor area is needed, too, and that, not withstanding any extra costs for stronger walls, should have a beneficial effect on building costs. This may be further enhanced by the fact that self feed bunkers are often roofed over, and this is certainly unnecessary when mechanically feeding. It is even possible to use a non-walled clamp.
- Flexibility in the siting of silos in relation to animal housing is enhanced, because silage can be transported.
- Several different groups of stock can be served from one silo.
- Silage can be weighed and rationed out with reasonable accuracy.
- Opportunity is afforded to incorporate silage with other feeds before it is fed.
- There is potentially less wastage by comparison to self feed silage.
- There is more even intake of silage by stock along a feed barrier or manger of adequate length. Bullying is also less of a problem as compared to self feeding.
- There are potentially fewer slurry problems, including less time and effort spent scraping concrete where the feeding area is immediately next to winter housing and possibly arranged all under one roof. With self feed the area to be scraped increases as the silage face recedes.

Choice of unloading and feeding system

In designing an appropriate unloading and feeding system, due regard has to be taken of a number of factors including the distance from silo to stock, the number of animals and quantities to be fed. Whether feeding is done on a once or twice daily basis, or every few days, will determine the consequent capacity and work rate required. To minimise labour time input the emphasis must be to make every trip count and to handle as large amounts as possible, minimising the number of loads to be taken to cattle during the feeding session.

Care is necessary in the selection of silage handling and feeding equipment to ensure that it is entirely appropriate for the particular farm circumstances in question. This may appear obvious but there are many instances when mistakes are made. In particular, it is vital that equipment will adequately accommodate building layout and access, width of feeding passages and height of feed troughs, and that it has sufficient reach related to height of the silage clamp. Further factors include ability to cope with variations in the physical form of silage, such as chop length and moisture content, and the possible necessity to mix silage with other feeds before being fed. The means of accurately gauging feed quantities, preferably by weighing, is yet another consideration and although this might significantly inflate capital costs, weighing is the key to accurate rationing. Whatever equipment is selected it must be sturdy, reliable and simple to maintain – not least for the fact that stock persons are not normally famed for their mechanical prowess! Finally, capacity of fodder boxes or mixer wagons should be large enough to minimise the number of separate journeys necessary to expedite the feeding of stock.

Silage unloading and feeding equipment

There are three categories of equipment used:

- Silage unloaders.
- Silage unloader-feeders.
- Silage feeders.

Silage unloaders

FORELOADERS
Tractor mounted hydraulic foreloaders are in use on most farms, providing a cheap means of extracting silage at rates up to about 7 min/tonne, while this can certainly be reduced to 4 min/tonne or less with specialised materials handling loaders. Conventional forks leave an untidy surface more prone to allow air ingress between loading periods with consequent deterioration and loader tine spacing is not always suitable for short

chopped silage. A power operated loader can help leave a smoother face by raking the silage downwards and then picking it up off the floor. But a relatively inexpensive silage grab is better, obviating any need for pre-cutting of flail or double chop materials which can be extracted straight out of the silo.

There are two main versions of grab in popular use. Tined grabs comprise a horizontal fork with close tine spacing (around 165–200 mm) to reduce spillage of short chopped grass and a pivoted impaler crowding top fork served by a pair of double acting hydraulic rams. Here the downward action of the crowding grab ensures a clean face and full load. A wide range of sizes are available to suit most loaders and tractors and materials handlers, with capacity from 500 kg to around 1 tonne.

Shear grabs have been introduced more recently where the impaler tined fork is replaced by a sturdy pivoted knife blade arrangement cutting through the silage on three sides to extract a block in as little as 15–20 seconds. Shear grabs are now one of the most popular systems of extracting silage from clamps, combining simplicity with facility to minimise waste. As well as speed, the major advantage here is that the cutting action seals the silage face of the clamp and the block itself, which will help to reduce deterioration. Furthermore, blocks can be left to stand for a few days without undue risk of secondary fermentation, which can enhance flexibility in feeding arrangements.

Whilst shear grabs are used not only to unload but also to transport blocks to the point of feeding, normally over shorter distances, it is usual for grab unloaders to be used to fill some form of forage box or feeder.

Loaders should have sufficient reach not only for clamp height (maybe 3–4.5 m) but also for filling feeder wagons. A bunker not less than 9 m wide is necessary to allow adequate room for manoeuvring equipment. Adequate counterbalance weights are needed to create tractor stability, and torque converter transmissions, with instant forward/reverse, will help reduce wear and tear on the clutch and gearbox. Specialised materials handler vehicles really come into their own in these circumstances and their heavy duty hydraulic system and larger lift capacity can rule out the need for other transport where the distance is short and a suitable feed barrier is arranged. But here there is little opportunity for rationing and some hard work is often inevitable to avoid wasting silage beyond the reach of stock.

SLEW LOADERS

This type of loader, fitted with a suitable grab and operated from a stationary tractor, involves less manoeuvring and can accommodate a narrower bunker silo. They are capable of achieving good work rates (typically 4–5 man min/tonne), with far less wear and tear on a conventional tractor. However, by comparison to the performance of more recently introduced equipment, slew loaders do tend to produce more of a ragged working

Plate 4.1 Tined silage grab (*Chillton*).

Plate 4.2 Silage shear grab (*Parmiter*).

surface conducive to ingress of air and likelihood of deterioration between successive unloading periods.

PUSH OFF BUCKRAKES

Tractor rear mounted push off buckrakes fitted with short tines have been used to extract full chop or double chopped silage and transport it to the feed area. Pre-cutting of the silage with knife or saw may be necessary and certainly so with long material. This practice together with its limited ability to deal with higher clamp faces, unless the buckrake is fitted on to a rear mid lift or foreloader, which limits its capacity, has led to the push off buckrake now being a relatively unpopular means of unloading and moving silage.

TRACTOR MOUNTED BLOCK CUTTERS

Block cutters were introduced into the UK in the mid 1970s from continental Europe. Blocks of silage are cut from the silo leaving a clean cut clamp face and producing a 'sealed' block that can be fed over several days without deterioration. Other particular advantages are that chop length and dry matter are not particularly important, and the block provides a sound basis for volumetric assessment of quantity of silage fed. Cut blocks range in size from around 0.85 to 2.5 m^3 (300–1000 kg), with cutting time taking 0.5–3 minutes. For larger units the block cutter is tractor rear linkage mounted, which limits the height of operation to about 2.5 m, but a reach of up to 4.5 m or so is achievable with front mounting on present generation telescopic boom type farm material handlers.

Once extracted, blocks are normally taken and deposited straight to the feeding area, either at the feed barrier, in a ring or rectangular feeder, or taken to outlying stock for self feeding maybe over several days. However, output is very much influenced by distance between silo and stock, and transporting blocks in this way is generally inappropriate for larger herds. The current generation of block unloaders also incorporate a conveyor feed distribution system, described later. Alternatively, blocks can be deposited with care into appropriate feeder boxes or wagons which are strong enough to cope with compact blocks of short chopped material, although if this is regular practice, this will put considerable stress on the feeding mechanism.

There are two main categories of cutting mechanism based on either a travelling reciprocating knife or a single action shear blade system. With reciprocating knife cutters a hydraulically driven single knife or saw may reciprocate in a vertical plane and is driven round a horizontal rail with rounded corners, around three sides of the block. Other systems employ horizontally reciprocating cutters on two or three sides of the block which are moved downwards hydraulically, or a cutting chain on three sides is operated hydraulically. Sometimes electronic controls are offered as an option. In some cases remote electronic controls are used in the tractor

Plate 4.3 Typical block unloader (*Audureau*).

cab to operate electronic solenoid valves controlling hydraulic oil for both reciprocating and lateral movements.

A holding down device may also be fitted to allow the knife to operate in firm material for a clean cut and to assist in stable transport. In all cases, the cut block is supported on tines which may also incorporate some form of push-off mechanism or tilt ram system for transportation and offloading of the block. Reciprocating type block cutters employ a relatively complex hydraulic drive system, adding to the capital cost and maintenance needs. These types are now tending to become superseded by much simpler shear grab unloaders, previously referred to under foreloader unloaders. These are not only less expensive but are simpler to operate and are capable of extracting blocks more quickly. Blocks are cut by the action of a pivoting heavy duty front and side shear blade cutter which slices through the silage in a single action process powered by a pair of double acting hydraulic rams. A wide range of models are available with capacities from around

Fig. 4.1 Rotomix 3500 combined cutter unloader mixer wagon.

300 to 900 kg, the latter of which is appropriate for industrial farm and telescopic material handlers.

CUTTER LOADERS

Cutter loaders have been mainly used in continental Europe and have created little interest in the UK, primarily because there are cheaper alternatives available. A number of different versions have been developed, but all have the basic configuration of a cutting head moving down the clamp face, loosening silage which is collected and elevated into a feeding vehicle by a conveyor. Some have used a rotating cutter moving through an arc which can 'work' a face up to 4.5 m high for fitting on to a light tractor. Some use belt conveyors while one version has used a pto powered blower. One self propelled version combining mixer wagon is shown in Plate 4.4.

Silage unloader-feeders

SELF LOADER-FEEDERS

These are of two types: those that fill a feeder box with loose silage from the silo face and those that cut whole blocks of silage. With both types silage is removed with the minimum of face disturbance then transported and fed to stock along a feed barrier by side discharge conveyor.

Loose silage self loader feeders are well established, particularly in continental Europe. They are usually rear mounted or semi mounted, holding 2–3 m^3 of loose silage, although versions are available in trailed form which hold up to 4 m^3. Bulk density of loaded silage is around 250–350 kg/m^3.

The feed box is reversed up to the silage face and with semi mounted and trailed machines the back end is lowered to the ground into the filling mode by hydraulically raising the transport wheels. With the mounted

Plate 4.4 Mounted self loading feeder box (*Multimate*).

version the box is tipped hydraulically into an appropriate angle to receive silage from the feed face.

There are two main means of filling. In an earlier model silage is cut by means of rotating cutters on the rear of the feeder as the unit is raised and lowered, and material is thrown into the transport hopper. This cutting mechanism has been generally succeeded by a more robust and simpler hydraulically operated comb or rake which is moved successively down the silage face until sufficient silage has been gathered. The comb/rake is pivoted and controlled hydraulically on the end of twin or single, and sometimes telescopic, arms which also move in an arc by means of another arm. The angle of the teeth can be altered hydraulically with one model. The effective use of this type of unloader-feeder is determined by the degree of operator skill in achieving maximum fill and leaving a tidy face.

Feed delivery is by means of a heavy duty chain and slat conveyor which takes the silage across the feeder box to a toothed drum that shreds it into a readily acceptable condition, delivering to the manger or feed barrier. In one trailed self loader-feeder silage is moved forward by chain and slat conveyor, shredded by two or three rotating toothed drums and delivered to a side delivery chopper/thrower for discharge at any height up to 2 m. Both the filling comb and discharge mechanism are hydraulically powered with infinite control of delivery output independent of the forward speed of the tractor. Although right hand delivery is the norm, left hand delivery is available on special order.

These boxes can handle a range of other feeds including brewers' grains, beet pulp, maize and potatoes. Some of the larger trailed versions will

Plate 4.5 Multipurpose self loading feeder (*Lucas*).

also accommodate big bale silage provided that, when loading, bales are broken up, and they can also double up to chop and blow straw for bedding up to a distance of 15 m using a discharge hood that is controlled hydraulically from the tractor seat.

Block unloader feeders are a more recent development. Typically, the block is extracted by a hydraulically powered shear grab, as described

Plate 4.6 Shearblock unloader-feeder (*Montec*).

earlier, then tipped into the feeder box by hydraulic ram operated pivoting tines. Except for front mounted versions, a further block can then be extracted and carried 'in reserve' on the tines, then tipped into the box once the original has been discharged. As with loose silage feeders, discharge is achieved by chain and slat conveyor and shredding rotor driven by hydraulic motors, the speed of which in infinitely variable.

To simplify the hydraulic services and control required some feeders use remote controlled electronic solenoid operated hydraulic valves to divert oil to the appropriate ram or motor.

Both types of unloader-feeder suit buildings with limited access and are particularly suitable for smaller herds, perhaps up to 80–100 cows. Cycle times of 6–9 minutes can be achieved, with the block unloader-feeder marginally giving the best potential performance and neatest cut face, but this type provides less versatility for other uses.

Silage feeders

Though dual purpose unloader-feeders have an obvious appeal, in a number of circumstances it is often more appropriate to feed cattle with separate equipment including:

• Self feed trailers.
• Tractor mounted forage feeders.

- Adapted or dual purpose FYM spreaders.
- Purpose built forage box trailers.
- Feed mixer wagons.

SELF FEED TRAILERS

Though less commonly used now and generally for the smaller herd, these provide a low cost means of both transporting silage and serving as a feed manger. There are two popular versions, one of which comprises a two-wheeled chassis trailer with a high level V-shaped rack from which animals can feed. The other is based on a converted standard low loading trailer fitted all round with a tombstone, or similar, barrier through which cattle place their necks to feed. It has been popular practice for some farmers to build one to their own design, and certain long wheelbase versions can accommodate up to 40 cows at once. As adequate all round access is needed in the feeding area these self feed trailers are not readily accommodated into many more conventional cattle building and yard layouts.

CONVERTED/DUAL PURPOSE FYM SPREADERS

A number of FYM spreaders readily convert for forage feeding and provide a particularly low cost means of distributing not only silage but also other bulk feeds separately, or in the same consignment as silage. In the simplest case, a conventional rear delivery chain and slat spreader can be converted for delivering along relatively narrow feed passageways where stock can reach most of the feed between the barriers, or a tractor mounted scraper-plough arrangement can be employed to push silage towards the barriers. However, it is more convenient to fit a variable speed cross conveyor, driven hydraulically or mechanically, normally at the rear of the spreader, for more uniform distribution of feed along the feed barrier with wider feeding passages.

A number of barrel type FYM spreaders can readily be converted for silage feeding by fitting a chute or baffle to deflect feed into the trough or feeder. At least two FYM spreaders are capable of feeding silage without conversion. With both, material is discharged sideways by means of a large pto driven, rotating, front mounted rotor with a deflecting chute controlling the height of delivery. With one, material is moved forward to the rotor by a conventional chain and slat conveyor, the other uses a moving tailboard which slowly pushes the entire mass of material forward towards the rotor. Inevitably, the major drawback with such dual purpose equipment is the considerable time and inconvenience involved in pressure washing the spreader when changing use from manure spreading to silage feeding. Where the manure spreader is to be used on a frequent and regular basis through the winter, other feeding equipment will need to be used.

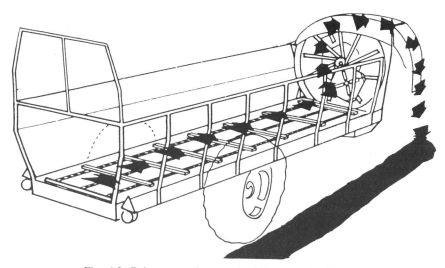

Fig. 4.2 Colman muck spreader/silage feeder (*Western*).

FORAGE BOXES

These are either tractor mounted or trailed self unloading boxes specifically used for feeding silage alone or a mixture of bulky foods. Some have a dual purpose role and can collect grass from the field, for silage or zero grazing, either from a forage harvester or directly from the swath as a self loading trailer.

Tractor mounted versions, suitable for the smaller enterprise, have a capacity of 1.5–1.8 m³ and some of them are capable of holding, shredding and delivering silage blocks. Trailed forage boxes generally have a greater capacity but greater filling height than spreader conversions and provide a more uniform discharge. Unloading is achieved by a chain and slat or belt floor conveyor moving material forward to two, or on some three, shredding rotors which tease out silage on to a cross conveyor for delivery to mangers or a feed barrier. Cheaper versions use open rotors of angle iron fitted with relatively long tines. This type is more prone to wrap long material than 'beater drum' rotors, which are fitted with relatively short tines. Sides of the feeder box are usually tapered, thereby reducing drag and releasing the load as it is carried forward by the floor conveyor. This may be a chain and slat design or belt type, both with adjustable speed control. The former gives a more positive and 'aggressive' feed of material on to the shredding rotors, whereas a load of relatively 'solid' material may tend to slip on a belt conveyor which may be seen as an advantage as a safety device or a disadvantage in the terms of limiting positive shredding action and consistency of output.

Cross conveyors are of three types: chain and slat, auger and flat belt. By comparison to belt conveyors, auger cross conveyors are cheaper and

Plate 4.7 Typical side delivery forage box using belt conveyors (*Kidd*).

simpler to maintain and they also provide a little more mixing action as feed is conveyed. However, their throw is limited, even if speeded up. Belt conveyors are driven by a hydraulic motor which not only enables speed, and hence throw, to be adjusted readily, but gives the opportunity to reverse conveyor direction from the tractor seat, too, by simply switching a directional control valve providing feeding to either side. This is of distinct benefit when access to the building is limited. The rate of discharge, often around 0.5 tonnes/minute, is largely dependent upon a combination of speeds of the main and cross conveyor, while the speed of the latter also affects the throw of the feed, which can be up to 2 m with some feeders.

Box capacities are in the range 4.5–10 m^3, holding up to 5 tonnes of silage, with delivery heights varying over 0.75–1.2 m depending on the model and whether an elevator is fitted. Some boxes can be factory fitted with chassis height extensions to facilitate delivery of silage over high walls or tombstone barriers. This extra height does not do much, however, for stability. Versatility of some forage boxes is further enhanced by the fitting of on-board load cell weighing facilities similar to those used for mixer wagons. Indeed, when so equipped, forage wagons often provide a cheaper substitute for mixer wagons for complete diet feeding of a combination of bulky foods using the admittedly limited mixing ability of the shredding rotors before discharge. In order to produce even moderate mixing of a typical maximum limit of two or three materials, it is necessary to arrange materials in the box with fibrous materials, including hay or straw, on the bottom and other materials, including concentrates, sandwiched between layers of silage. A fair amount of hand levelling in the box is necessary to achieve acceptable uniformity. To aid start up and delivery, materials should also be filled progressively towards the shredding drums.

Further refinements include a root door and concentrate hopper. Fodder roots, such as mangolds, are 'chipped' into manageable bite-size pieces on one version by the fitting of a hinged door which is swung down into a vertical position just in front of the shredding rotors, limiting flow of the roots to an opening about one-third of the depth of the fodder box. A concentrate hopper enables concentrates and other free flowing materials to be fed on to the cross conveyor, which further extends flexibility of these forage boxes.

In an ADAS survey on complete diet feeders, forage boxes produced an acceptable mix except at the end of unloading runs when there was, at times, a greater proportion of silage. To compensate for this some operators reversed the feed dispensing circuit on alternate days.

FEED MIXER WAGONS

Following the work of J.B. Owen in the mid 1970s which demonstrated the benefits of ad lib complete diet feeding of cattle, considerable interest continues to be shown in this system. This has been further emphasised by a growing awareness of the technique of flat rate feeding of concentrates to dairy cows and a trend towards a greater reliance on high quality bulky foods to replace amounts of concentrates fed. However, feed mixer wagons are not always used purely for feeding complete (balanced) diets as on many farms it may be desired to use a mixture of bulky and perhaps some concentrate feed in addition to some other form of feeding – such as concentrates via in or out of parlour feeders.

Feed mixer wagons are capable of mixing a range of materials with very widely differing properties simultaneously to produce a homogeneous mix and then deliver the mixed feed to stock. Ingredients may include silage, chopped or ground hay or straw, root crops, molasses and other food by-products and cereals. This would be time consuming to feed individually, requiring a number of operations. Some mixers will safely handle up to 10% of the load in the form of long hay or straw, but mixing is less efficient than with shorter material.

Mixer wagons offer a number of advantages compared to forage boxes. In the first place, mixing is sufficiently thorough to prevent animals having much success in selecting particular ingredients – apart from larger particles such as whole roots, unless they are chopped before loading. Secondly, bulk density can be enhanced with auger type mixer wagons, but this advantage is largely minimised where straw is fed. Bulk density is very much related to chop length and amount of fibrous material: thus a forage box might, typically, be capable of a bulk density of 250 kg/km^3, whereas with some mixer wagons this would be 300 kg/m^3 and sometimes even up to 350 kg/m^3. A further advantage in favour of mixer wagons is that they are sturdier and can handle effectively a more comprehensive range of materials. There are three main types in common use:

- Multi auger type.
- Paddle type.
- Chain and slat type.

Multi auger type
This was the original type of mixer wagon introduced from the USA, and still remains popular. Here one, but more usually three or more, longitudinal augers rotates slowly within a V-shaped body. A popular version uses a formation of two top augers and one in the bottom of the V rotating in the opposite direction so as to mix the feed longitudinally. Augers have a high power requirement: a tractor capable of not less than 45 kW at the pto is needed, and to minimise high starting loads augers should be rotated slowly throughout the filling process.

Mixed feed is discharged through a side hatch more conventionally sited near the front of the body and delivered by a chain and slat conveyor with hinged extension to facilitate entry to a building and also to control delivery height. Feed output is controlled by the height of an adjustable hatch shutter.

Plate 4.8 Auger type mixer wagon (*Waterson Engineering*).

Plate 4.9 Paddle type mixer wagon with auger discharge (*Keenan Easi Feeder*).

Paddle type

The mixing function is performed by a large, sturdy, slowly rotating paddle arrangement carried on a central shaft through a U-shaped hopper.
Feed is discharged in one of two ways, either directly at a high level where
a hydraulically adjustable canopy shutter is opened up along the top edge
down one side or indirectly using an auger. With the latter type, once
mixing has taken place a long guillotine shutter is raised along one side of
the hopper which enables the paddles to feed the entire length a large
longitudinal discharge auger. The auger has the capacity to provide very
fast and complete delivery of material through a hatch, with adjustable
canopy and tray, near the front of the machine at the left hand side. When
the guillotine gate is raised hydraulically a linkage also opens the auger
discharge canopy. The mixing discharge height is commonly around 1.1 m.

This type of feeder has developed a particular reputation as a sturdy,
reliable, simple mixing system with low maintenance costs, providing fast
emptying using auger discharge. The latter aspect is not only important
in respect to time saved, but also because some feeders may require the
selection of a tractor gear lower than is provided in order to discharge an
appropriate volume of feed along the feed barrier or manger. However,
in mixing the paddle mechanism does not provide a positive distribution
of feed along the hopper and care is needed in loading, standing the machine on level ground and adding a pre-mix of concentrates before loading
roughages. Inevitably, the mixing action produces a typically light loose

material, of lower bulk density than for most auger mixers, hence the overall weight per unit volume capacity is reduced.

Chain and slat mixer wagon

This type is not in common use. Here mixing is achieved by a chain and horizontal slat conveyor traversing a circuit of the base, front, top and rear of the wagon hopper. All mixer wagons are normally fitted with on-board load cell weighing facilities to weigh in individual feed ingredients and to weigh out the mixed feed. Accuracy is normally in the range of ±2% and can be checked against a pallet load of fertiliser or feed. It is important to remember that accuracy is enhanced as the load increases, thus smaller quantities of concentrates are best added after the bulky feed. A fair degree of skill is needed by the foreloader operator to ensure that pre-determined quantities are not overshot when bucketfuls are discharged. Not only is a visual and audible warning system necessary to warn of load levels, but a 'traffic light' early warning system will further help the tractor driver.

Organisation of loading is optimum when all bulky ingredients are within a reasonable distance of the silage clamp. Concentrates can be loaded by auger, from pre-loaded bins, or more usually from foreloader bucket or from pre-loaded hoppers or reusable big bags.

Mixer wagons range in capacity over $4-12$ m^3. With a bulk density of $250-300$ kg/m^3 and feeding, say, 45 kg/cow per day, a typical 10 m^3 mixer wagon should hold enough to feed 55–66 cows in one load. Labour requirement for loading, mixing and distribution is around one person hour per 100 cows per day. In theory, several times that number could be fed with one machine, raising the possibility of group sharing, although this is likely to be fraught with considerable logistics and disease risk problems. Mixer wagons are generally reckoned to be economic for herds of 150 cows and over.

Organisation of feeding clamp silage

As in any materials handling situation, the whole system should be integrated and planned to provide smooth organisation and effective use of time. Thus, sufficient space is needed for either U or reverse turns (15 m turning circle), long straight through vehicle circuits with adequate width and height for entry and exit and ideally distances should be kept to the minimum. Passageways should be wide enough to allow feeders to unload with minimum waste and to avoid hand work. Overall work rate performances will vary considerably from holding to holding, depending on individual circumstances including operator skill and motivation. As a guide, however, Table 4.1 indicates typical performance for various clamp unloading and feeding systems. Larger capacity equipment and tight operating routines will enhance performance, while poor layouts, the

Plate 4.10 This simple tractor rear mounted plough device enables 'nosed out' silage to be speedily pushed back to the feed barrier (*Sila-Push, Hareland Engineering*).

extent to which additional ingredients are to be added and the number of separate operations will reduce performance. Any hand work will certainly reduce output considerably. Thus, follow up hand forking or distribution of silage along the feeder barrier should be avoided if at all possible. Feed outside the eating zone of stock can be moved nearer using a tractor mounted blade, brush or other attachment.

Table 4.1   Typical unloading and feeding performance for clamp silage.

System	Guide to loading and feeding rate (person minutes/tonne)
Foreloader and forage box	8–11
Slew loader and forage box	7– 9
Foreloader and mixer wagon (complete diet feeding)	12–14
Foreloader direct to feed point	10–18
Block unloader and feeder Loose silage self loader/feeder	9–15

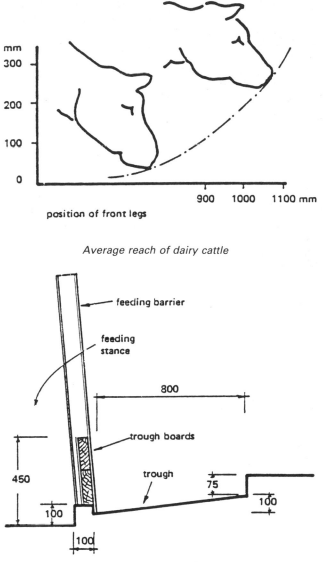

Average reach of dairy cattle

Detail of the trough incline

Fig. 4.3 Detail of feed barrier arrangement (*Cermak*) (Centre for Rural Building, Aberdeen).

Especial care is needed with feed trough design to ensure that cattle feed intake is not restricted, that waste is minimised and at the same time that the need for hand work is curtailed. For mature cattle, with ad lib feeding, trough/feed barrier length can be as little as 200 mm, otherwise when feeding at specific times an access width of 700 mm per animal must be allowed. This can be reduced to 450 mm for 6-month old cattle, increasing to 570 mm for 18-month old cattle. At floor level, feed located further than about 850 mm from the stance will be beyond the reach of most adult cattle. Figure 4.3 indicates a suitable feed barrier arrangement to contain feed and minimise handwork. Alternatively, a wall can be built along the front of the trough to retain feed, the height of the wall allowing clearance for the feed wagon discharge chute.

Barriers must be sturdy enough to restrain stock of all ages, prevent bullying and avoid injury to stock from any feeding system. Barriers constructed from vertical tombstones, individual yokes or horizontal, vertical or diagonal rails can all be used successfully.

Accuracy of rationing is very much facilitated when feed boxes are fitted with strain gauge weighing facilities and pre-set load signalling. Where these are not fitted, pressure-sensing gauges tapped into the hydraulic system (such as the 'Wayload' system) provide a low cost means to assess the weight of each forkfull. Success is influenced by regular calibration and care in gauge interpretation. An earlier survey indicated that if carefully used, this system can achieve accuracy with loading silage of the order of ±6%. However, of course, it is left to the operator to total up the quantities accurately while loading continues!

Big bale silage

Big bale silage now represents a significant proportion of conserved fodder in certain situations. These include fodder for the smaller herd, in the upland situation, and a spontaneous means of overcoming overflow from a clamp system. Ideally, a dry matter of 30–40% will minimise the weight and number of bales required as well as effluent problems, but around 25% is more usual. Bales are usually 1.2 m wide and 1.2 m in diameter, weighing around 0.5 tonne each with a density of 300–350 kg/m^3. This is about half what can be achieved with a clamp silo, so emphasis has to be placed on achieving anaerobic conditions as fast as possible by tightly wrapping them or inserting them in a close fitting bag and ensuring that birds, vermin or anything else do not cause any punctures. More recently, the techique of bale wrapping using stretch film has streamlined the sealing operation and further enhanced the popularity of big bale silage in some areas.

In the smaller scale situation, big bales are often self fed either by loading into a ring feeder or by placing bales side by side along a feed barrier and feeding over several days. However, it has to be remembered that the higher the dry matter of silage, the less stable the bale will remain once open. Attachments are available for foreloaders which incorporate hydraulically operated tines/grips for holding a heavy bale vertically downwards to allow precise handling.

Where bales are to be broken up for feeding the long, unchopped nature of the material does cause problems for equipment that chops and lacerates silage as it is distributed. A more successful approach may be to unroll the bale, either along the ground by simply using the tractor foreloader or by using a mechanical unroller. There are at least three devices in use. One comprises a simple foreloader attachment using a pair of freely rotating spiked plates to grip the bale each side along its central axis and allow it to be conveniently unrolled along the ground. Hydraulically driven spike bale spinners are also available for front or rear tractor mounting. With the rear mounted version, once the bale has been speared the spear axis is swung hydraulically through 90° to give either left or right hand delivery. These spinners are normally used for unrolling bales of straw for bedding which some can achieve within 20 seconds – much too fast for silage. However, with judicial use of hydraulic controls a fair job of unrolling and unravelling can be made, although it is inevitable that uniformity of distribution will not be ideal. Orientation of the spike is important in achieving uniform unwrapping.

Alternatively, unrolling can be achieved using a side delivery moving bed unroller of the type show in Plate 4.11. Here the bale sits in a cradle and is turned by a spiked conveyor chain running down one side and across the bottom. As the bale turns in the cradle, silage is delivered over the side of the unit. Driven by a reversible hydraulic motor and mounted on the

Plate 4.11 Round bale roll out feeder.

tractor's three-point linkage, the feeder can discharge over barriers of around 600 mm or into feed troughs. Adjustable protruding forks give further control over bale rotation. A particular advantage of this type of machine is its ability to prevent wastage during transport, particulary where shorter grass is involved.

A tub grinder can also be used to chop big bale silage to lengths of around 30–150 mm. One bale can be chopped in about 6 minutes using a 45 kW tractor, with the chopped material discharged by high or low blower chute to a trailer or feed passage. A big bale tub chopper, as used for distributing straw into cow cubicles, has been used fairly successfully for feeding big bale silage.

Tower silage feeding

Haylage towers, for the storage of high dry matter silage, were introduced into the UK from the USA in the late 1950s, but have drastically declined in popularity in recent years. They combine the potential for the conservation and storage of particularly high quality silage with the labour saving features of a highly mechanised feeding system using forage boxes or a fully automatic system based on fixed conveyors. Of course, fully automatic systems provide obvious attractions – in removal of the tractor and driver requirement as well as diversion of time saved towards more effective herd management control. With conveyor feeding, too, there is scope to economise on building space needs in respect of feed passage widths, turning spaces and headroom.

To exponents of tower silage systems, however, the most attractive aspect has been the ready means of conserving and feeding a higher quality high dry matter product than can often be achieved in clamp silos, by exploiting the considerable tower consolidation effect and reliable means of tower sealing to encourage rapid anaerobic conditions. In store, losses can be as low as 2.5–8.00% DM, compared to 15–20% with some bunker systems.

Despite these advantages, investment costs are inevitably high and earlier equipment did tend to be somewhat unreliable, although later equipment has been improved. A fixed conveyor feeding system is also less flexible, particularly so when breakdowns occur and alternative means of feeding have to be introduced. It is vital to make some alternative feeding system available if the breakdown lasts for more than a few hours.

Tower unloading

To minimise potential problems in unloading from towers, material should be reasonably consistent in chopped length and dry matter and uniformly distributed in filling to give even consolidation. Secondary fermentation is minimised if a minimum of about 75 mm is removed each day. With

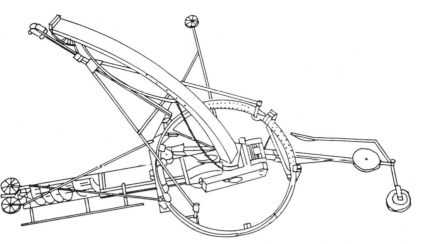

Fig. 4.4 Tower top unloader (*Boythorpe*).

concrete/fibre cement sheeted towers a top unloading system is normally used, operating best with material in the range 35–40% DM while bottom unloading is typically associated with steel sheeted towers which can accommodate material in the 40–45% DM range.

Though free access is easier for top unloaders, it is inevitable that last material in is first out, which is not always appropriate, and it is not possible to add more crop to a top unloading tower already in use. Whilst bottom loaders give a first in/first out system, they do need to be more robust than top unloaders and are thus more expensive, but they can be removed and employed to serve more than one tower. Furthermore, bottom unloaders compare favourably to top unloaders by avoiding the need to climb up the silo in order to adjust the unloader chute.

Top unloaders are of several types, the rotating auger being the most common. Silage is teased out by a slowly rotating large diameter auger incorporating knives which convey material to the centre where an impeller sucks it up and blows it down the outside delivery chute. In one version it is necessary to go up the tower to adjust the angle of the chute after about 1.5 m of silage has been used. In another version a cable and winch system is used to lower progressively a large diameter ring gear system which forms a circular rack for rotating the unloader auger round the tower. In another form of top unloader, the rotating auger is replaced by a three-arm rotor carrying a kind of notched 'disc plough' assembly which scoops material to the central impeller.

For bottom unloading a rotating arm and endless cutting chain unloader system is used. This chain arrangement is inserted into the bottom of the tower via a trench beneath the lower floor. The unloader comprises two parts: a sturdy endless chain conveyor working in the trench conveying

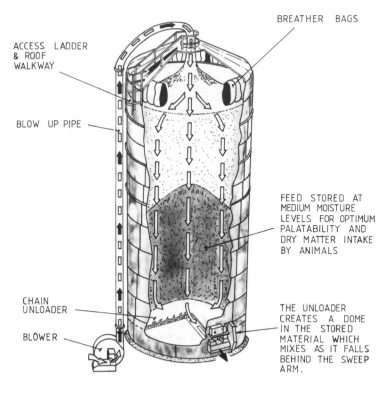

BREATHER BAGS

ACCESS LADDER
& ROOF
WALKWAY

BLOW UP PIPE

FEED STORED AT
MEDIUM MOISTURE
LEVELS FOR OPTIMUM
PALATABILITY AND
DRY MATTER INTAKE
BY ANIMALS

CHAIN
UNLOADER

THE UNLOADER
CREATES A DOME
IN THE STORED
MATERIAL WHICH
MIXES AS IT FALLS
BEHIND THE SWEEP
ARM.

BLOWER

SCHEMATIC DRAWING SHOWING PRINCIPAL FEATURES
OF HARVESTORE SYSTEM. THE ENTIRE STRUCTURE IS
MADE FROM RIGID STEEL, COATED ON BOTH SIDES WITH
FUSED LAYERS OF GLASS. BREATHER BAGS IN TOP
COMPENSATE FOR CHANGES IN AIR PRESSURE WHILE
MINIMISING CONTACT OF AIR WITH FEED. UNLOADING
IS VIA ATLAS OR GOLIATH CHAIN UNLOADER.

Fig. 4.5 Tower bottom unloader (*Permastore*).

silage from the centre to the outside and a cutting chain. Driven from the
end of the conveyor chain, this cutting chain rotates around the floor
delivering cut material to the centre for outloading by the conveyor chain.
Insertion of the unloader is facilitated by a layer of chopped straw in the
bottom of the tower. To minimise secondary fermentation it is advisable to
keep the unloading hatch closed between unloading sessions.

The unloading rate from both top and bottom unloaders is somewhat
variable according to dry matter, density variations and, in the case of top
unloaders, the rate at which the winch is operated to lower the unloader
as unloading continues. Electronic controls are sometimes employed to
monitor the current consumed by a top unloader, which can then control a

reversing motor winch to raise or lower the loader as conditions indicate, and in such a way minimise the possibilities of motor stalling. Monitoring of the electrical current consumption over the time span of a complete revolution will determine whether the unloader needs raising or lowering for the subsequent revolution.

The most effective means of controlling quantities fed is by the use of a continuous flow weigher. Such a weigher can also be used to record the amount of silage loaded into the tower.

Feeding from tower silos

Silage can be delivered to the feeding situation by one of three methods: forage box, pneumatic conveyor or fixed conveyor.

Forage box conveying and feeding certainly does provide the greatest flexibility in that the tower need not be adjacent to the feed area, other feeds can be readily included and it may be possible to deal more easily with breakdowns. However, not only is the bonus of full automation lost, but the system has to rely on the availability of tractors and drivers.

Occasionally, the forage blower used for filling the tower can be used once again for blowing silage to the feeding area up to 100 m away. To achieve this, flexible ducting, around 450–600 mm in diameter, is linked to a cyclone from which feed drops into the trough.

Mechanical fixed conveying is much more usual, using either distribution or progressive feeders. With distribution feeders, almost immediately the conveyor system is started some feed is available over the length of the trough, so that cattle do not crowd to one end. Examples include chain and slat, chain and flight and most belt feeders. Progressive feeders first fill the manger at the feed end, supplying the far end when the remainder of the manger is full. Such feeders require that stock are shut away until filling is completed. This can be easy enough to achieve with a dairy herd as feeding is carried out during milking. Examples of the progressive type include open augers and jog trough feeders.

Chain and slat conveyors, whilst of simple design and ability to cope with a wide range of materials, are prone to wrap long material and to wear from abrasive materials. One type consists of a chain and slat mechanism running over a tapered bed. As material is dragged along it is gradually dropped from the tapering edge into the feed trough beneath.

Endless belt feeders are the most common type of conveyor feeder remaining in use, with a lower power requirement and few moving parts, and thus they tend to be more reliable. However, they are prone to leakage of silage under the belt causing wrapping and also belt slippage. For this reason, most belts are driven by open rollers to avoid the build up that can occur on solid rollers. Belts are commonly 300–400 mm wide, running on rollers at about 30–90 m/minute. Feed is diverted off the belt to mangers by an angled plough or rotary brush which is winched to and fro above the belt. A plough will commonly be towed at about one-tenth belt speed, while the rotary brush is operated at about half belt speed. By reversing the direction of the rotating brush, feed can be diverted to either side of the belt. Belt conveyors are particularly useful where concentrates are to be added to silage for complete diet feeding. By fitting an integral load cell belt support system a continuous weighing facility can be built into the belt feeding system.

Open auger feeders comprise a pair of vertical boards with an auger mounted between them. The auger is adjusted in height above the floor of the manger to accommodate the pile of feed which builds up beneath and progressively from one end. A number of fully or semi enclosed auger systems have also been used including one where the auger rotates in a cylinder with a series of slots. In the latter, the cylinder is initially positioned with the slots uppermost while the auger fills up, then the cylinder is rotated to move the slots to the bottom allowing feed to be delivered at a reasonably constant rate. Other auger feeders make use of half round U- or C-shaped casings. On start up, delivery commences at one end but soon develops along the whole length.

Jog trough feeders comprise a long suspended trough system which is oscillated by a crank mechanism to move material progressively along the trough.

Mechanised feeding of hay and straw

In the handling and delivery of hay or straw to stock the emphasis has to be to minimise the amount of manual work where possible and reduce waste. Each farm will justify its own bale handling system which is normally based on tractor and trailers or, better, the farm all terrain materials handler such as the telescopic boom type. These are particularly useful for loading big round bales into circular cattle feeders. Where hay and straw is fed in racks, waste can be reduced when the rack is lined with weldmesh to minimise the amount of feed that stock can pull out at one time.

Where bales are to be fed along a feed barrier, large round bales can be unrolled using either a bale spinner or bale unroller – as previously described for big bale silage. For conventional sized hay or straw bales, use could be made of a straw bale chopper towed along the feeding area with its bale hopper filled manually. Such choppers are commonly driven by a single cylinder engine and mounted on a three-wheel trolley which is pulled along by the stockperson. Care needs to be exercised to ensure that chopped material is not unduly wasted by stock. The adjustable delivery hood can be set to allow material to be thrown some distance for bedding or deflected downwards for feeding. Big bale tub choppers can similarly be used not only for single big bales, but some will even accommodate ten rectangular conventional bales.

Plate 4.12 Straw bale chopper (*Taylor Fosse*).

Straw for incorporating into mixed rations should be chopped or ground, although fine grinding does tend to minimise the roughage element and upset feed intake. Ammonia treatment of straw enhances feed value and increases digestibility. Treatment is achieved either by the oven method or by treating straw in plastic bags with anhydrous ammonia. Once treated, straw is stored until needed and fed from the bale.

Feeding of relatively free flowing bulky feeds

The handling and feeding of relatively free flowing bulky feeds such as brewers' grains, stock feed potatoes and swedes is generally quite straightforward. They can either be incorporated with silage and fed with forage or mixer wagons or fed separately. For the latter some silage feeders can be adapted, otherwise a number of simple tractor mounted fodder feeding boxes are available. These comprise a three-point linkage or fore end loader mounted hopper which is tipped for filling. A single or double

Plate 4.13 Tractor mounted fodder box (*BOM*).

longitudinal auger, driven by hydraulic motor, delivers material to one or both sides. A 'cutter wheel' can be attached on the delivery outlet to enable root crops to be chopped up.

References

ADAS (1983) *Mechanical Feeding of Silage with Other Feeds.* Booklet 2148.
ADAS (1984) *Mechanised Unloading and Feeding of Clamp Silage – an Equipment Guide.* Welsh Office, Agriculture Department.
ADAS (1987) *Big Bale Silage*, P3096.
Amos, G.E. (1984) 'Feeding out – from self-feeding to full mechanisation.' *British Grassland Society Occasional Symposium No 17*, 51–67.
Buriak, P. and Walker, P. (1986) 'Horizontal auger type and drum type feed mixer.' *Journal of the American Society of Agricultural Engineers* **2**(2), 97–100.
Catt, W.R. (1980) 'Mobile equipment for dispensing forages.' *British Society of Animal Production Occasional Publication No 2*, 147–154.
Cermak, J.P. (1977) 'Some ergonomic considerations of mobile silage feeding systems.' *Farm Building R & D Studies* **9**, 1–5.
Davidson, W. (1984) 'Fodder feeding.' *Agricultural Engineer* **39**(2), 61–66.
Kay, M. (1984) 'The ruminant's requirements.' *Agricultural Engineer* **39**(2), 58–60.
Raymond, F., Redman, P. and Waltham, R. (1986) *Forage Conservation and Feeding.* Farming Press.
Redman, P.L. (1980) 'Unloading forages from store.' *British Society of Animal Production Occasional Publication No 2*, 125–133.

5 Mechanised cattle concentrate feeding

Introduction

This chapter examines the efficient, accurate and hygienic mechanised feeding of concentrates to dairy cows, beef cattle and calves. The financial value of consumed concentrates represents a highly significant proportion of variable costs for dairy herds. Thus, for instance, as the margin of milk production value over concentrate consumption costs is considered one of the premier indices when measuring the efficiency of a dairy enterprise, the mechanisation aspects of moving and rationing concentrates to the cow obviously merit close scrutiny.

Unlike bulky feeds, the feeding of concentrate feeds presents perhaps a greater challenge in respect to the speed, frequency and number of individual, relatively small parcels of feed which must be delivered, in variable quantities, to stock and most often in the execution of another process – milking. Although considerable studies have been made in developments of feed control systems, thus easing the boredom of the stockperson, it is only relatively recently that improved versions of the actual feeding dispensers have started to become available with more acceptable standards of accuracy.

Not only can physical bulk density affect accuracy of dispensing rations, but feeder design, maintenance and frequency of calibration are perhaps the most critical determinants in feeding economy. The introduction of milk quotas has particularly highlighted the importance of controlling costs. A number of surveys have demonstrated quite clearly that, for the majority of dairy enterprises, there is much scope for improvement in accuracy of feeding. Over a 300-day lactation the saving of 100g/cow per day for a 100 cow herd represents 10 kg/day or 3 tonnes feed per annum amounting to £400–£500. In some instances, savings of up to 1 kg/cow per day might be achieved at the peak of lactation, and on an annual basis the savings can be very significant indeed. It is evident that close attention to feed rationing can thus pay handsome dividends.

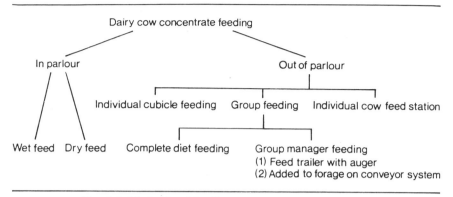

Fig. 5.1 Methods of feeding concentrates to dairy cows.

Table 5.1 Bulk density requirements of feedingstuffs

Form of feed	Bulk density (m^3/tonne)
13 mm cubes	1.56
10 mm pencils	1.73
Meal	1.95

Concentrate feeding to dairy cows

Figure 5.1 summarises the main means of feeding concentrates to dairy cows practised in the UK.

In-parlour concentrate feeding

General considerations

(1) QUANTITIES OF FEED AND SPACE REQUIREMENTS

Quantities consumed are much dependent upon yields and feed policy. Total consumption could thus be in the range 0.9–2.4 tonnes/cow per annum, with the greatest demand occurring during the peak calving period outside the grazing season. Thus, a 100-cow Autumn calving herd with average yield 6500 kg might mean an average herd peak daily need of perhaps 600–800 kg of concentrate split into two equal feeds of 300–400 kg. If concentrates are bought in 20-tonne lorry loads one consignment will thus last a maximum of 25–33 days. Table 5.1 gives a guide as to bulk density and space requirement for concentrate feeds. Concentrate storage facilities should accommodate a full delivery load with a suitable reserve to accommodate delays in supply.

(2) STORAGE AND MOVEMENT OF CONCENTRATE FEED

Hoppers serving parlour feeders will need to be kept topped up from an adjacent bulk supply, either from an overhead loft or conveyed into the parlour from an outside free standing storage bin or feed store room. An overhead loft can provide a relatively cost effective and space economic means of storage with the added probable bonus of abolishing the need for mechanised conveying, with bought-in concentrate cubes blown into the loft space from the delivery lorry and hoppers beneath filled by gravity. The loft floor should not only be designed and constructed to carry safely the weight of concentrate involved (typically 500–700 kg/m^2), but also preferably designed to minimise hand shovelling when the supply is beginning to dwindle. Here a hoppered loft floor presents a distinct advantage over a flat floor.

A free standing bin does, however, offer a number of advantages over loft storage. Not only is the parlour simpler to construct, but overhead ventilation and natural lighting are improved. There is no problem of loft sealing to keep dust out of the parlour and water vapour out of stored feed. Feed levels in an outside bin can be checked visually through a transparent panel or electrically by diaphragm or proximity switch. Alternatively, by mounting the bin on load cells, overall food consumption can be monitored continuously. Details of suitable bins are given in Chapter 2.

For bulk supplied, bought-in concentrate feeds, filling of bins or lofts is simply effected via the delivery lorry's pneumatic conveyor. Delivery pipes should be 100 mm in diameter, with an exhaust pipe of not less than 150 mm diameter. Even distribution of feed in feed lofts can be facilitated by using a multi outlet delivery tube system. Bends in delivery pipes should be at least six times the pipe diameter, and all joints need to be taped or sealed to prevent dust emission. Sleeve joints need to be arranged so that fast moving cubes do not get broken against protruding edges.

Home produced rations may be transported by mobile feed trailer with auger or pneumatic discharge. Alternatively, where feed preparation facilities are within around 100 m, a fixed conveyor system should be appropriate.

Parlour feed dispenser hoppers are filled from outside storage bins normally by rigid or centreless auger or by chain and disc conveyors. Rigid augers are popular, but to save on the number of separate drive motors needed, whenever a change of direction occurs, special right angle corner drive versions are available. Most augers are capable of handling cubes up to 13 mm in diameter, the degree of breakage being influenced by auger diameter, number of corners, speed of rotation, the number of drop-off points to be negotiated and, equally important, the strength and stability of the cubes.

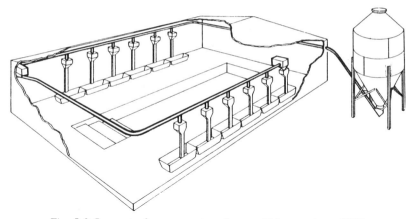

Fig. 5.2 Layout of auger system for a milking parlour (*EB*).

Flexible augers are now the most common means of conveying in milking parlours, with the advantage of minimising the number of drive units and with the capability of accommodating bends of not less than 1.5 m radius and distances up to 60 m. Alternatively, chain and disc conveyors, although not suitable for cattle cake, are appropriate for conveying meal and can turn on a very tight radius, conveying vertically when required and forming a continuous circuit round the parlour.

(3) LIMITATIONS ON QUANTITIES CONSUMED

The relatively short period that cows occupy the milking parlour certainly places a limitation on the amount of concentrates that can be consumed. This limit, the 'available concentrate eating time', will be influenced by the form of feed supplied to cows, the parlour type and work routine, milk yield and milk flow time.

Whilst there will be variations between cows, research shows that dry meal can be consumed at around 0.25–0.33 kg/min, 13 mm cubes at 0.33–0.50 kg/min and 5–6 mm pencils at around 0.40–0.55 kg/min, whilst liquid feed gives consumption at a dry food equivalent rate of 1–1.3 kg/min or more. Manger shape and position and the total quantity of feed presented to the cow do have an influence. Table 5.2 summarises data compiled by the Scottish Agricultural Colleges.

In a herringbone parlour the available concentrate eating time will depend upon whether each stall is supplied with its own unit or if it has to share it with the opposite stall. In the latter case, provided cow identification and feeding are carried out whilst the cow is entering the parlour, the available concentrate eating time is twice the unit time (UT) – that is, she can continue to feed whilst the cow opposite is milked before the unit is transferred to her. The actual amount of feed time available is summarised in Table 5.3.

Table 5.2 Dairy cow concentrate consumption rates.

Type of concentrate	Dry matter (%)	Eating rate (kg/min)	Time to eat (3.5 kg/min)
Dry meal	86	0.32	10.9
13 mm cubes	86	0.37	9.5
Rolled moist barley	81	0.44	8.0
5 mm pencils	86	0.47	7.4

Table 5.3 Available concentrate eating time in herringbone parlour

No of stalls/unit	Available feed time
2	2 × unit time – batch change time
1	unit time – batch change time

The unit time is the time it takes to extract milk from that cow (the milking out time) plus the machine idle time. Tables 5.4–5.6 summarise typical values for these.

Table 5.4 Milking out times for typical average yields.

Average milk yield for that milking (kg)	Milking out time (min)
10	4.8
15	5.9
18	6.5
20	6.9
22	7.3
24	7.7

Derived from NIRD data where milking out time $(t) = 0.207x + 2.75$ min, where x is the average milk yield in litres.

Table 5.5 Machine idle time.

No of stalls/unit	Machine idle time
2	about 0.2 min
1	(0.8 × work routine time) × n/2
	where n = No of units

Table 5.6 Machine idle time and batch change times.

Parlour type	Machine idle time	Batch change time (min)
10 × 5	0.2	1.0
10 × 10	4.0	1.0
12 × 6	0.2	1.2
12 × 12	4.8	1.2
16 × 8	0.2	1.6
16 × 16	6.4	1.6
20 × 10	0.2	2.0
20 × 20	8.0	2.0

As an example, to calculate the available feed time (AFT), where the average yield is 20 litres and the work routine time is 1 minute:

milking out time = $0.207x + 2.75 = 6.89$ minutes.

In a 12 × 6 parlour, with a machine idle time of 0.2 minutes, the calculation is thus:

Unit time (UT)　　= milking out time + machine idle time
　　　　　　　　= 6.89　　　　　　+ 0.2　　　　　　= 7.09 minutes
Available feed time = (2 × UT)　− batch change time (BCT)
　　　　　　　　= (2 × 7.09) − 1.2　　　　　　　= 12.98 minutes

Therefore the maximum feed that can be consumed is approximately 4.8 kg of 13 mm cubes. In a 12 × 12 parlour with a machine idle time of 4.8 minutes, the calculation becomes:

Unit time (UT) = milking out time + machine idle time
　　　　　　　= 6.89　　　　　　+ 4.8　　　　　　= 11.69 minutes

AFT   = UT − BCT
                     = 11.69 − 1.2   = 10.49 minutes

Hence the maximum feed that can be consumed is approximately 3.9 kg of 13 mm cubes.

In practice, it is unwise to plan to feed more than 3.5 kg per cow at any milking. Problems inevitably ensue where newly calved cows, on a lead feed programme, may require 5–6 kg concentrates at each milking, necessitating an available feed time of around 13.5–16 minutes. Alternative feeding arrangements, such as out of parlour feeding, effectively come into their own in these circumstances.

Parlour feed dispensers

(1) DESIGN FACTORS

With the introduction of milk production quotas, many dairy producers have been persuaded to reduce the concentrate content of the ration. It is thus even more relevant that concentrate dispensers work accurately and effectively. Underfeeding will not only reduce milk yield but may also influence body condition and have wider physiological implications for the cow. Overfeeding, on the other hand, can lead to wastage, possibly cause digestive upsets, and might slow the work routine in the parlour. The cost of overfeeding where dispensers may be 20% over generous can easily exceed £50/cow per annum at current prices. Accurate feeding will help animals to perform to predictable levels of production and will contain production costs. Where equipment is to be used at least twice a day throughout most of the year, operators need to be confident that it will dispense feed both quickly and reliably and to high standards of accuracy. All too often, field surveys of parlour feed dispensers have indicated widespread inaccuracy, sometimes of alarming proportions, due not only to poor design and installation but also to lack of adequate care and attention from operators in respect to maintenance, calibration and adjustment. In a survey carried out by the Scottish Agricultural Colleges, 8% of feeders had errors ±50% or more, 37% had errors greater than ±20% and only 35% were within ±10% of target weight.

In striving for high standards of accuracy it should be remembered that the higher this is, the higher the potential cost of equipment. But it is relevant to define more clearly what is meant by accuracy in this context. Accuracy is a measure of the closeness achieved to the desired target and is composed of two distinct factors: 'precision' and 'bias', also referred to respectively as 'repeatability' and 'offset'. Thus, it is desirable that the dispenser should consistently dispense units of feed over a period of time with little variation in amount dispensed, that is good precision or repeatability, and ideally the amount dispensed each time should be very near to the target value, that is low bias or offset. In practice, high precision and low bias tend to be expensive to achieve. The emphasis for any mechanism should certainly be towards an acceptable standard of repeatability, and a good dispenser will have a coefficient of variation of less than $\pm2\%$ on a delivery of 5 kg compound feed. This means that when the output is measured over a large number of times, about two-thirds of the samples would be within $\pm2\%$ of the mean. The mean could well differ from the target but this difference, the offset, should be capable of being adjusted to zero or allowed for in the feed calculation.

The main factors affecting offset in volumetric feeders are the physical properties of the material, including particle size, moisture content, angle of repose, coefficient of friction, and so on, as well as the mechanical features of the dispenser, for example operating speed and vibrating effects on the mechanism. Thus, as the speed of rope or handle pull often has a significant effect on the consistency of amount of feed dispensed with many manual feeders, there are clear benefits in favour of mechanised feed dispensing.

The most appropriate choice of feeder will depend on many factors, not least on parlour type and make. The propensity towards 'brand loyalty' often determines that feeders selected will be of the same make as the rest of the milking installation. But the fundamental selection parameters should be cost, accuracy, reliability, ease of calibration and adjustment and type of control.

An expensive feeder does not necessarily guarantee accuracy, as much of the higher cost may be generated simply because of a better and more advanced control system. Often such a feeder may still utilise the same type of inherently inaccurate dispensing mechanism as used in less sophisticated feeders. A gravimetric dispensing mechanism is not only fundamentally more accurate than most volumetric dispensers, but should also require less frequent calibration and adjustment. But as most feeders on the market are of the volumetric type it is best to opt for one with a near positive displacement dispensing mechanism, with little risk of carry over between batches nor danger of further amounts being delivered when the feeder is knocked. Some older feeders of the flat moving plate type, where feed is retained on a shelf by its natural angle of repose, are notorious for the common practice of intelligent cows soon determining that banging the side of the drop tube causes more feed to trickle over the edge of the dispensing plate. Such feeders are best situated outside the parlour with feed entering feed mangers from dispensing drop tubes through holes in the wall.

Field studies show significant variations in bulk density with compounded dairy feed and, in one case study with compounded meal, there was a variation of $\pm10\%$ of the mean. Thus, a certain amount of inaccuracy has to be accepted with volumetric feeders, but this can be minimised by high standards of calibration and setting. This will only occur where design permits the mechanism to be quickly and simply adjusted, with the minimum of fuss and need for hand tools. Hence, a single positive handwheel mounted externally on the side of the feeder may be better than a number of stepped adjustments which may be difficult to make.

Some electrically operated dispensers have a particular advantage in the provision of a simplifed compensation for bulk density. Thus, following a new delivery of feed a sample is weighed at just one feeder, then all the other feeders can be adjusted by a single control adjustment. Any adjustment system should enable whole units of delivered feed to be simply and directly correlated to the method by which feed levels are selected. Mental conversion factors should not be necessary.

Dispensers are actuated mechanically by Bowden cable or rotary shaft, or operated pneumatically by piston or bellows or by electric low voltage motor. Whereas cable and rotary shaft systems are mostly confined to manual control, pneumatic and electric actuators are more appropriate for semi automatic and automatic systems. Pneumatic actuating systems are relatively simple and make use of the parlour milking vacuum system. However, with the trend to drive more and more equipment by this means, a check needs to be made that the vacuum pump capacity is large enough to ensure a good vacuum reserve, which will help to minimise vacuum fluctuations. In a similar way, control systems may be operated pneumatically, but are more commonly electrically/electronically based. However, the damp conditions prevailing in milking parlours are not necessarily conducive to the proper working of electrical components. Thus, it is essential that boxes housing these components are absolutely watertight and sited in a position where accidental hosing can be avoided when the parlour is washed down. A 12 V or 24 V DC electrical supply is normally used for safety reasons.

All control dials or keypads should be clearly marked and calibrated and designed to enable rapid and precise settings to be achieved.

Dust and bridging problems are two more design factors to take into account. Inevitably, dust will be a problem with non-cubed concentrates and can be alleviated by fitting lids or covers to dispenser hoppers where appropriate, by angling the trough so that material slows down by friction rather than by impact and possibly modifying the drop tube. However, where baffles are fitted in the latter there is the possible danger of causing blockages. Feeding cubes or pencils greatly reduce dust problems.

Bridging of feed in storage bins, feed lofts and dispenser hoppers is much aggravated by damp atmospheric conditions and condensation drips. Again, avoidance of meal based concentrates will help, but use of steep sided hoppers and low friction linings will also help, as will elimination of constrictions. Good housekeeping by regular cleaning, of course, reduces the problem. However, where difficulties persist some form of mechanical or electrically vibrated agitation system may be necessary.

By contrast to where feed is supplied from an overhead loft, installations where dispense hoppers are filled directly from an overhead conveyor tend to have few bridging problems.

There are good arguments to locate feeders outside the parlour with gravity feed through holes in the wall. Problems associated with dampness are reduced, dust and noise minimised and cow interference with the dispenser is eliminated.

(2) TYPES OF DRY FEED DISPENSING SYSTEMS

Figure 5.3 summarises the most common dispensing systems.

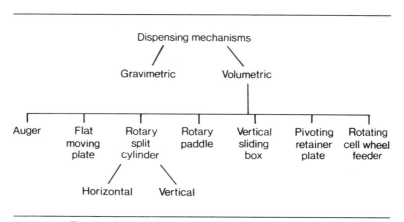

Fig. 5.3 Dry feed parlour dispensing mechanisms.

Gravimetric dispensers

Over the years, in emphasising the inherent inaccuracy of volumetric dispensing mechanisms, field surveys have prompted a certain amount of developmental work on mechanisms that dispense by weight. In the mid-1970s the then National Institute of Agricultural Engineering (now AFRC Engineering, Silsoe) developed three promising types of gravimetric feeder based on units of feed being weighed out in a bucket supported by a strain gauged arm, with models developed for static and rotary parlours. Other developmental work on gravimetric feeders was also undertaken at Writtle College. Various attempts were made by companies to launch a commercial gravimetric feeder and, sadly, the Gravimetric Feeder was withdrawn from the market within a few months due to technical problems. In the mid-1980s a new type of gravimetric feeder was launched: the Weighmaster (Fullwood), as shown in Fig. 5.4.

Feed is delivered from the hopper through an adjustable feed gate by means of an auger system and drops into a counterbalanced weigh tray. The weigh tray comprises a multisegment, intermittently rotating tray assembly which, when any tray is filled to the appropriate weight, over-

FRONT VIEW

Low voltage electric motor

Auger & Paddle

SIDE VIEW

Auger & Paddle

Stop holds tray for filling

Counter-weight

An auger draws feed from both ends, delivering to the centre where a paddle pushes feed forward – dropping it into a segment of the weigh tray.

When the weight of the feed equals that of the pre-set counter weight the balance tray tips, emptying feed into chute which takes it to the manger.

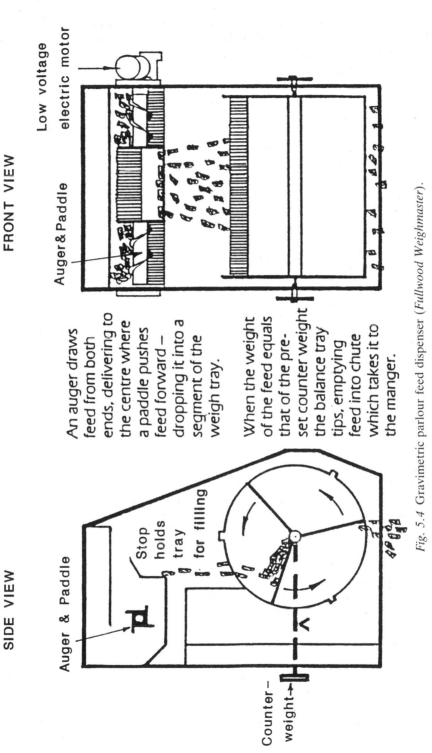

Fig. 5.4 Gravimetric parlour feed dispenser (*Fullwood Weighmaster*).

comes an adjustable counterweight beam and tips slightly, enabling a fixed stop to be cleared and allowing the tray to rotate by one segment so releasing feed down the drop tube. Once feed has been released, the tray resets itself and returns to its fixed state. In tipping about its pivot point the side balance arms, which retain both the rotating trays and counterbalance beam, bring a magnet close to a proximity switch which causes the auger to stop temporarily each time and thus avoid overshoot. At the same time, one increment of feed is deducted from the total feed requirement entered into the control system.

The normal setting for a portion is 500 g, but some adjustment can be effected by positioning a sliding weight on the left hand balance arm, sliding it up or down on a rod and locking it in position. The throat restrictor and feed lip also have some effect on calibration.

Gravimetric dispensing not only gives considerable potential for enhanced accuracy, but also reduces the time requirement for calibration and adjustment. Thus, it is estimated 5–10 person hours per annum may be required for gravimetric feeders in a ten-stall parlour in comparison to an estimated 43 person hours per annum for volumetric feeders, which may take 10 minutes each to calibrate and adjust every fortnight.

It is anticipated that more types of gravimetric feeder will enter the market in the next few years.

Volumetric dispensers

Auger dispensers – auger feed discharge is used in a number of feeders, not only for in parlour feeders but also it is the normal dispensing system for out of parlour feeders. In principle, a parallel or stepped auger moves material from the feed dispenser hopper to the drop tube, being driven either pneumatically or more commonly electrically. Some feeders incorporate twin augers which move feed in opposite directions to drop tubes serving two mangers with independent control of amounts dispensed to each.

As a semi positive displacement system, augers are perhaps one of the better types of volumetric dispensing mechanism with the slight possible disadvantage of the potential to damage cubed concentrates. In some systems, should the unit be knocked, carry over or dribble may be minimised by using a type of exit ramp.

One type of auger feeder is driven pneumatically from the parlour milking vacuum system which drives a piston and cylinder assembly. The latter is served by a pulsed vacuum supply from a pulsator, fed off the main vacuum pipe, and operated through a relay by a 12–15 V DC source from the central control box, which might be calibrated in 1–10 units for each feeder.

A sprocket with free wheel device, mounted on the auger shaft, is rotated via a short roller chain coupled to a connecting rod and attached

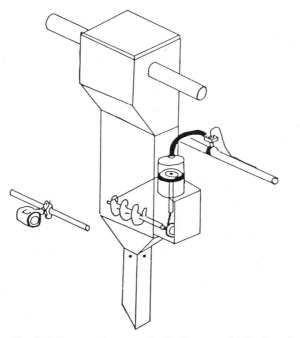

Fig. 5.5 Pneumatic auger feed dispenser (*Alfa Laval*).

to the piston, as shown in Fig. 5.5. When subjected to vacuum the piston is drawn upwards and the spring returns the piston when the vacuum is released, with the chain freewheeling over the auger drive sprocket. Adjustment of the amount dispensed is achieved by altering the piston/ connecting rod stroke length by means of an adjustable stop.

A number of suppliers use a 12–15 V DC electric motor to drive an auger system either via a connecting rod system or by direct drive. For the former a sprocket, roller chain and connecting rod system is used in a similar fashion to the pneumatic drive. This time the connecting rod is driven by a low speed, high torque car windscreen motor-gearbox unit. An adjustable slide on the side of the hopper alters the size of auger aperture, thus regulating the amount of feed dispensed.

A rather simpler arrangement is achieved with direct coupling using a similar geared down 12 V DC electric motor. In this system a much wider range of amounts can be dispensed, merely depending on the duration of motor rotation, in comparison to other systems which deliver a specific number of fixed portions. Following calibration checks, the amount dispensed can be relatively simply adjusted by compensating for the length of running time or the rate of delivery via an adjustable slide controlling the amount of feed entering the auger.

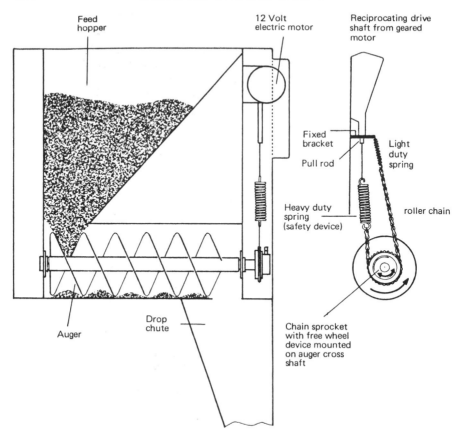

Fig. 5.6 Electrically operated auger feed dispenser.

Flat moving plate dispenser – this has been in popular use for many years. Concentrate feed from the hopper is retained on a horizontal base plate and is prevented from falling from the edge by its natural angle of repose. Delivery is effected by a pair of hinged plates which normally form a 90° angle horizontally along the top surface and to the base vertically to the rear. Feed is delivered when the hinged plates are pivoted to form a 180° angle which then coerces feed to fall over the edge down the drop tube. A cross shaft, which causes the pivoting action, is connected by a ball crack to a flexible rubber bellows which, with the aid of a return spring, produces a reciprocating action when alternately subjected to vacuum and atmospheric pressure.

Adjustments to the unit amount discharged each time are achieved by a screw stop limiting the movement of the hinged plates, and also the angle of the horizontal plate is adjustable.

In the past this type of feeder has been very prone to knocks from cows,

Plate 5.1 Electrically operated auger feed dispenser (*ATL*) (showing installation in parlour).

causing further feed to leak over the edge of the horizontal retaining plate. This problem has now been largely overcome by the fitting of a nylon brush arrangement across the feed discharge point. The brush helps retain feed, but also allows for it to be pushed forward under the action of the moving plate.

Rotary split cylinder – This type has also been in popular use for many years. It comprises a horizontal open topped cylinder or trough which is supplied with feed from the hopper above. When the trough is rotated through 180° feed is delivered to the manger beneath and at the same time the circular base of the inverted trough prevents the discharge of further feed from above. The trough is returned to the original rest position by a strong return spring. The unit is driven via a pulley from a Bowden cable system, either hand or mechanically operated.

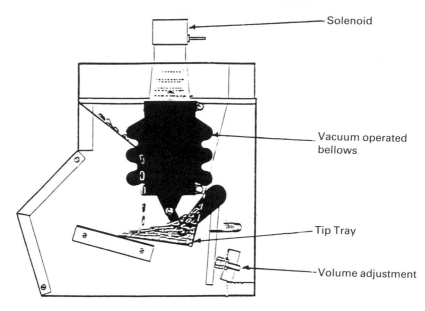

Solenoid

Vacuum operated
bellows

Tip Tray

Volume adjustment

Feed falls into tip tray which is attached to the
drawer linkage, after a set time the solenoid
activates the bellows.

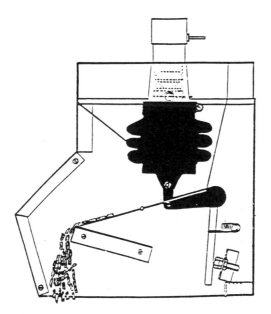

The tip tray straightens emptying feed into the
chute which takes it to the manger.

Fig. 5.7 Flat moving plate feed dispenser (*Fullwood Ration Master*).

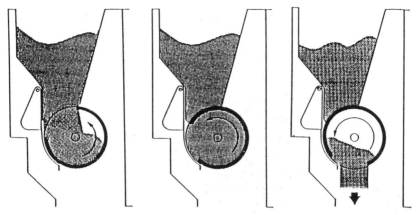

Fig. 5.8 Rotary cylinder feed dispenser (*Orby*).

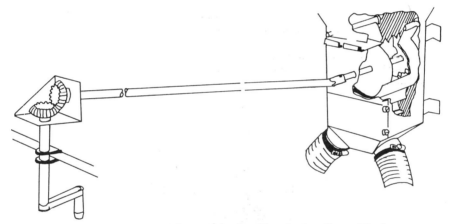

Fig. 5.9 Rotary paddle type feeder (*Rotafeeder, Peter Allen*).

One particular advantage of this type of dispenser is that a defined volume of feed is 'captured' and conveyed through the dispensing mechanism – a feature which assists towards improved standards of accuracy. The actual volume of feed discharged may be adjusted by rotating a hand wheel on the outside of the dispenser unit, so moving a slide to vary the internal volume of the split cylinder. This method of adjustment is simple, precise and easily effected.

Another version of the split cylinder dispenser uses a vertical hollow cylinder supplied from a hopper above. Feed is delivered when the cylinder slides round sealing off the feed point and allowing concentrates to drop through an orifice to the delivery chute. There is no adjustment to the volume delivered at each operation but two sizes of cylinder are available from the manufacturer.

Rotary paddle feed dispenser – this simple type of mechanism is normally supplied for manual operation, although a pneumatically operated version is now available. Feed is removed from the hopper and delivered by a simple four-bladed paddle wheel which, in rotating, traps a specific volume of feed between the blades and the curved paddle housing, before releasing the unit of feed at the bottom to the drop tube and then the manger. The paddle shaft is extended via a series of universal and right angle joints, terminating in a hand wheel for the dairy person to operate in the milking parlour pit. Alternatively, the unit can be fully mechanised by fitting a vacuum ram and cylinder assembly with ratchet freewheel device.

The amount of feed dispensed in each operation is adjusted externally by sliding a 'calibration plate' in or out of a slot. The effect of this is to reduce or increase the width of the paddle blades, which comprise two interlocking halves that can be partially telescoped together.

Vertically sliding box – this dispenser has relatively recently come on to the market. Feed, supplied from a supply hopper above, flows down a cranked drop tube to the manger beneath. A vertically sliding box arrangement is incorporated in the sloping middle section of the square section feed chute with strategically placed slots in each side. The box is attached

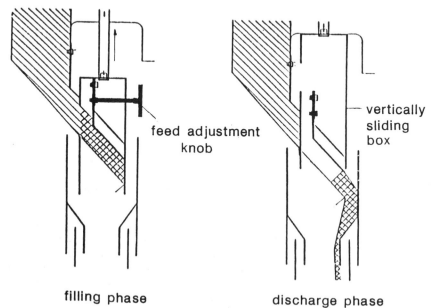

feed adjustment
knob

vertically
sliding
box

filling phase   discharge phase

Fig. 5.10 Vertically sliding box dispenser (*Quartermaster, Manus*).

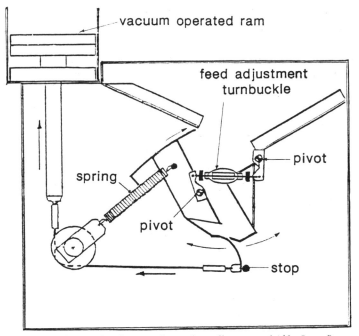

Fig. 5.11 Pivoting retainer plate dispenser (*Alfa Laval*).

to a connecting rod coupled to an overhead vacuum operated ram and cylinder assembly. When the box is in the lower position feed slides down the sloping chute, enters the box through a slot and is retained on the sloping floor. When the outer walls of the box are raised an exit slot on the other side of the box aligns with the retained feed, which is then released to the manger.

Calibration of the amount dispensed is via an external screw adjustment which controls the volume of feed retained during each operation. There is also an adjustable slide controlling the amount of feed entering the box.

Pivoting retainer plate dispenser – this type of feed dispenser is a relatively new entrant to the market. Here feed flowing from the supply hopper enters a small chamber comprising a pair of pivoting interlocking retainer plates held closed by a spring. The two plates are coupled by an adjustable turnbuckle linkage which adjusts the distance between them and thence the volume of feed retained between them. Feed is released when a vacuum supplied ram and cylinder assembly pulls the plates apart via a connecting rod and roller chain system. Spring pressure retracts the ram when the vacuum is disconnected and closes the plates. The top section of one of the retaining plates incorporates a curved blanking plate which, when the plates are pulled apart, closes off the feed supply orifice above.

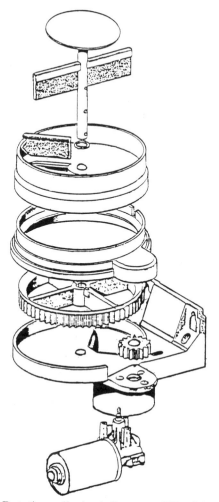

Fig. 5.12 Rotating cell wheel dispenser (*Westfalia Separator*).

Rotating cell wheel dispenser – this dispenser comprises a multisegmented feed dispensing disc incorporating teeth around its periphery and which is gear driven by a DC electric motor. This segmented disc rotates in a feed chamber which is filled from above through a slot in the base of the feed supply hopper and a rotating paddle and disc arrangement in the bottom of the hopper ensures that the segmented disc is adequately supplied. Feed trapped in each segment is conveyed round until it falls through a hole in the feed chamber floor to the manger beneath. There is no adjustment for the unit amount fed apart from changing the cell wheels, which provide either 60 g or 100 g concentrate feed per impulse of the feeder mechanism.

Other types of feed dispenser

Various other feed dispensing mechanisms remain in use, such as those based on paddles, dribble/vibrator and bottomless box, but they are now increasingly superseded by more recent developments.

CONTROL OF PARLOUR FEED DISPENSERS

Control of feed dispenser operation can be effected manually, for example by the number of turns or pulls of a handle, mechanically and pneumatically, using spring type sequence timers, but most usually nowadays electrically using a low voltage DC supply. There is a choice of three main systems of control: manual input, semi automatic 'memory' feeders and fully automatic systems.

Manual input

This still remains one of the simplest and most popular systems where the dairy person visually identifies each cow, refers to a feed chart and then pulls or turns a handle a certain number of times or, more usually, sets the appropriate ration on the selection switches or key pad. Often the control incorporates a selector switch for each pair of stalls and a change-over switch determining which side is to be fed. When the 'start' button is pressed, all the stalls on one side of the parlour are fed simultaneously. An alternative system, which does not require the dairy person to refer to a feed chart, is either to batch cows according to yield and flat rate feed or use colour tail bands, changed weekly or fortnightly according to yield. In this case all the dairy person has to do is move the selector dial to the appropriate colour segment corresponding to the current tail band colour. Because of the inconvenience of changing tail bands and the availability of improved equipment, this practice is now much less popular.

Semi automatic control

Over the years a number of semi automatic control systems have been developed, including a milk recorder jar weighing system enabling delivery of concentrates according to milk flow and another system using plastic punched cards, the number and spatial arrangement of holes corresponding to a particular quantity of feed. With the availability of cheaper and more sophisticated electronic equipment these systems have been fully superseded by so-called 'computer' or 'memory' feeders. The dairy person has still to identify cows but then has only to 'key in' the identities in stall sequence via a push botton keyboard and then press a 'feed' button. The unit automatically steps on until all the identities entered have been fed. This much reduces the mental effort required by the dairy person at milking and can speed up milking performance.

Built-in data recording and interrogation facilities also provide a valuable aid to herd management. Thus, a digital display gives a running total of the number of cows milked, total amount of feed dispensed and an

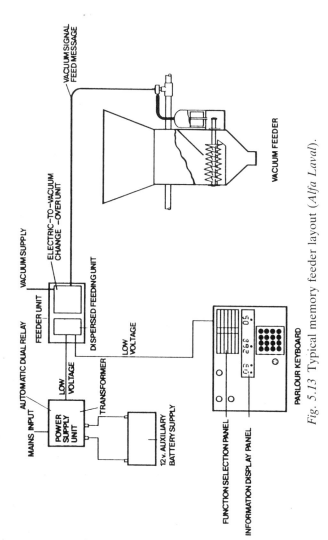

Fig. 5.13 Typical memory feeder layout (*Alfa Laval*).

'alarm cow' display facility reminds the operator to check on individual cows. Such a system particularly shows its worth in alleviating some of the potential problems associated with relief operators. Manual over-ride controls are provided so that cows could be fed in the event of an electronic failure or if, for some reason, it is necessary to over-ride the memory for specific ad hoc needs.

Fully automatic control

Here cows are automatically recognised as they enter the parlour and thus all the advantages of memory feeders are supplemented by the capacity for greater operator freedom and the potential for enhanced milking performance, with the capability to link up with an out of parlour feeding system as well. Such automatic feeding systems are often suitable for incorporating into a comprehensive automatic computerised dairy herd management system, including automatic milk recording, automatic adjustment of feed rates, full cow data recording and analysis and an 'alarm' signalling system. Thus, most auto recognition feeding systems provide means for operator on-line communication with the memory data bank system. A digital display and/or warning light system warns the operator of items needing attention for individual cows, such as 'AI check' or 'dump milk', while such information can be simply entered on to the parlour feeding console by the operator in abbreviated code form. Figure 5.14 shows a schematic layout of such a system.

Automatic cow recognition is achieved by the cow wearing either a passive transponder on a neck collar or an active (battery powered) transponder incorporated into an ear tag. When the transponder comes within range of an interrogator unit the transponder is energised by a radio frequency signal and its identification code is transmitted back and analysed by the control computer, thereby giving individual cow recognition. Inevitably, recognition of every cow is not always achieved, chiefly because the transponder is outside the range of the interrogator. American work at the University of Illinois suggests that the minimum acceptable rate of identification for most milking parlour installations is 98%. This work confirmed that the range of passive neck collar transponders is approximately 150 mm while that for active ear tag transponder is approximately 460 mm.

American trials over a period of 5 months showed rates of correct identification of 87.6% for neck band passive transponders and 93.5% for the ear tag system. The majority of missed identifications with the latter were caused by a few lost or malfunctioning transponders. Later work suggested that, with improved durability, the system would achieve the 98% goal.

Where a cow has not been recognised, a warning lamp should light up on the control unit indicating the stall position in order that the operator may enter the number manually. Battery life for active transponders is

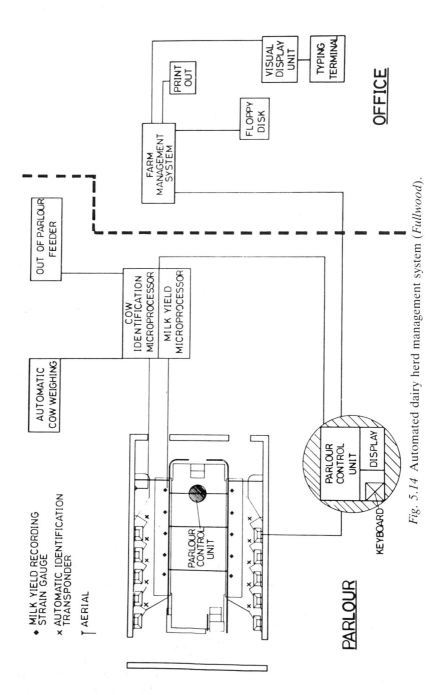

Fig. 5.14 Automated dairy herd management system (*Fullwood*).

expected to be at least 5 years with the unit queried a few times every day, but life expectancy decreases with increased frequency of interrogation.

MAINTENANCE OF PARLOUR FEEDERS

Survey work has demonstrated clearly that parlour feed dispensers are badly neglected on the majority of dairy enterprises. Whereas one would expect that as equipment gets older more regular checks are carried out, since the ageing process leads to wear and loss of precision, in practice the opposite is true. Only on a very small minority of installations are feeders calibrated as frequently as once a month, and in the Scottish Agricultural Colleges' survey, feeders received no calibration checks on around half of the installations. There is economic evidence to support routine checking of feeders fortnightly, or when changing batches of feed.

Calibration is quite simple to carry out and the only equipment needed is a bucket and an accurate weigh scale covering the range 0–10 kg, marked off in not more than 0.2 kg stages. It is suggested that around six recordings per dispenser should be made, comprising three readings at two different settings. It should be remembered that those dispensing mechanisms which have a higher coefficient of variation, such as moving plate dispensers, require a greater number of samples to achieve a high degree of confidence in the mean value obtained.

A periodic check needs to be made for gummed up deposits, bridging and for breakages and mechanical wear. Dampness and dust combine to form gummy deposits on surfaces which not only may impair accuracy over time but may cause total malfunction and bridging problems. Some installations are more prone than others to these problems. Whilst a periodic scraping out is a short term answer, longer term solutions need to be sought to minimise condensation and dust.

Checks need to be made for aspects such as loose, broken or missing springs, seized or worn auger bearings, loose, corroded or non-existent stops and poor electrical or vacuum supply connections.

Liquid feeding of dairy cows

It was established earlier that there is commonly insufficient time available during the milking routine for high yielding dairy cows to consume all the concentrates to which they may be entitled. Work carried out by the former National Institute for Research in Dairying some years ago demonstrated that by mixing meal in water at the rate of 400 g per litre, the consumption rate could be more than doubled for the dry material equivalent. Further advantages of wet feeding are that, as well as largely eliminating a dust problem, it does provide further encouragement towards use of home-mixed rations, avoiding the need for an expensive cubing process. However, the availability of out-of-parlour feeders and difficulties experienced in rationing and getting individual cows to accept liquid feed has meant that the practice has not been taken up to any wide extent.

The former NIRD wet feeding system employed off-the-shelf milking equipment components based on a milk receiver jar and milk pump assembly adapted as a continuous mixer fed with a calibrated supply of water and meal. Feed was rationed out using a vacuum operated ball-plug valve. Those few dairy producers who have adopted liquid feeding tend to use a modified pig pipeline feeding system with a continuous batch mixer. The extended availability of load cell weighing facilities incorporated into such equipment (see Chapter 3) helps to overcome some of the inherent problems in the accurate rationing of relatively small portions of liquid feed needed by dairy cows, as compared to larger amounts dispensed to large pens of pigs.

Out-of-parlour feeding

It was established earlier that, in most situations, constraints on time available limits parlour feeding of concentrates to a maximum of around 3.5 kg of feed per cow per visit. But in order to maximise yield potential it is important not to place any nutritional constraints upon the cow during the critical few weeks after calving, when peak yield will be reached. Indeed, work conducted by the former National Institute for Research in Dairying showed that 1 kg extra milk at peak yield is equivalent to an extra 200 kg milk throughout the lactation. It is inappropriate to consider slowing the milking routine to allow more time for concentrate consumption and, furthermore, most authorities accept that to avoid acidosis concentrate meals should be no greater than 3 kg. Thus, additional meals outside the parlour are likely to prove beneficial to any cows receiving more than 6 kg feed/day in two parlour visits, or 9 kg in three visits.

Benefits

The potential benefits from out-of-parlour feeding are considerable. Not only does it remove a constraint to maximum daily nutrient intake, but as feeding is organised on a little and often basis it can enhance feed conversion. Also, by ensuring more chemically stable conditions in the rumen, there is less risk of digestive upsets. A wider variety of feeds can be used which can both reduce costs and add variety to the diet. This, in turn, can stimulate appetite and maximise feed intake.

So how can supplementary feeding opportunities be organised? Perhaps the most obvious choice is to install out-of-parlour feed stations. Effective though these may be, they are by no means the only way of spreading the feeding of concentrates. There are a number of alternatives.

Where the need for supplementary feed time may be limited to a relatively short season, additional non-milking visits to the milking parlour

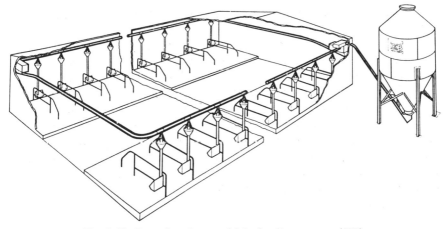

Fig. 5.15 Out-of-parlour cubicle feeding system (*EB*).

may be a low cost but tedious and time consuming solution. But barrier feeding of concentrates is a more popular practice with feed dispensed manually from sacks, by auger from a tractor mounted concentrate feed box or trailer, or mixed with forage and fed as a complete diet. The idea is attractive from a number of viewpoints: a wider variety of feeds can be fed including 'straights', exploiting the potential of opportunity buying, and also there is no requirement for cubed or pelleted feed which gives an incentive for home feed processing without the necessity for the expensive cubing process. However, there is no precision and guarantee that individual cows or heifers will consume their appropriate intake and timid animals are subject to bullying. The problem can only be alleviated by physical grouping, but this makes tedious extra work. Dispensing accuracy from auger type mobile feed trailers can be enhanced if the trailer is fitted with load cell weighing equipment.

Manger feeding in stalls or cubicles is yet another possibility and is a system popular in continential Europe. One company offers a cubicle feeding system utilising volumetric feed dispensers similar to those used for pig feeding. Concentrates are conveyed from an outside bulk bin by a centreless auger system to individual dispensers, fitted with adjustable telescopic drop tubes, in each stall in the cubicle house. An electric linear actuator operates a pull cable system to open feed dispenser doors allowing feed to pour down drop tubes into each manger. In this system the cows are usually housed in groups and all the feeders are set to deliver the same volume. The number of feeds per day can be automatically controlled by time clock and thus the ration can be delivered on a little and often basis, typically eight feeds a day. To ensure that all cows feed at the same time and to minimise bullying a bell rings for 45 seconds before feeding to allow animals to make their way and fill up all cubicles.

Out-of-parlour feed stations

Out-of-parlour feed stations are now a particularly popular alternative means of providing extra compound feed input for high yielding dairy cows. Automatic feed dispensing cubicles are strategically sited in or near the cubicle house, one to every 30 cows or so, from which privileged cows may obtain an appropriate quantity of feed, preferably in small amounts spaced over a 24-hour period. From very simple, rather unselective and generally inaccurate dispensers, they have developed into systems with the capability of allocating very exact amounts of feed to individual cows on a highly controlled basis.

Early out-of-parlour feeders were based on appropriate cows wearing either a metal neck chain or key to gain free access to feed, which was typically dispensed very slowly and/or with built-in delay periods to encourage cows to vacate the stall after a short period and so give adequate cow rotation. Studies showed that many of these systems failed to achieve their objective as some persistent cows could take more feed than economically justified and some failed to use the feeder adequately.

Current out-of-parlour feeders exploit the potential of the ubiquitous micro chip, mostly using the same type of auto recognition system as used for fully automatic parlour feeding and electronic sow feeding, both previously discussed. Cows wear passive transponders on neckbands or active ear tag transponders which enable each cow to be specifically recognised then fed to a carefully orchestrated computer controlled feed regime throughout each day and throughout the lactation, with on-going management feedback on feed consumption behaviour for each cow. Such feeders offer the height of convenience for outside parlour feeding, but at a price: currently around £60–£75/cow per annum at 1990 prices.

So what are the particular claimed benefits of out-of-parlour feed stations?

ADVANTAGES

Surveys carried out by the MMB indicated some of the following reasons given by dairy producers for purchasing out-of-parlour feed systems:

Saving of labour

This is particularly valid where a separate 'barrier feed' of concentrates can be avoided, and where automatic feed conveying and hopper replenishment are incorporated. Time may also be trimmed from milking, especially if parlour feeding is abandoned and the parlour is large enough to sustain higher milking performance. Labour hours saved do not necessarily translate into a lower labour bill, but time freed can be devoted towards better management control and stockmanship.

More accurate feeding

A number of popular parlour feed dispensing systems are, sadly, far from accurate and there is a consensus of opinion that out-of-parlour feed dispensers do generally provide enhanced accuracy. This is assisted to some degree by the fact that with the relatively small number of out-of-parlour feeders required, they do tend to be more regularly calibrated and better maintained than parlour dispensers. It seems likely, too, that the built-in resistance to the 'she deserves another pull today' syndrome is also another factor enhancing feed accuracy. So, too, is the equipment's ability to record and display the actual amounts dispensed: this stimulates the dairy person's on-going awareness of the need to control feed intake.

Little and often feeding

It is now a well accepted fact that feeding little and often ensures a more stable rumen pH and facilitates the more efficient conversion of feed.

Increased yields

Whilst some producers have recorded herd yield increases following the installation of out-of-parlour feeders, there is no emphatic evidence from surveys and trials to support the claim that increased yields can be expected per se. What is hard to prove is that on farms where yield performance was enhanced that the installation of out-of-parlour feeding was the trigger to overall management improvements. However, recent trial work has certainly shown that out-of-parlour feeding can enhance milk quality, for both fat and protein yield.

More feed to high yielders

For many dairy producers, this is the principal advantage. By permitting privileged high yielding cows the opportunity to maximise their concentrate intake, full milk yield potential can be obtained, but without the dangers of digestive upsets normlly associated with large intakes of cattle cake.

Easier management

The alternatives to out-of-parlour feed stations can present numerous management problems and stress. Thus, grouping animals into high and low yielders is not only time consuming but needs appropriate divisions in winter housing and general handling arrangements. Barrier feeding of concentrates again needs specific organisation and a certain time input, which also gives no guarantee on individual cow intake. Most importantly, microcomputer control of out-of-parlour feeding can enhance management by logging and analysing feed consumption and other management data, with a built-in alarm facility to identify cows which have not consumed their due amounts.

Physiological advantages

Supplementing those benefits of out-of-parlour feeding cited by producers, recent research work carried out by the AFRC Institute for Animal Health at Compton, Berkshire, has also demonstrated a number of physiological benefits. Researchers showed that out-of-parlour feeding provides benefits on blood glucose, total protein concentrates and body condition score accompanied by observed improved yields of milk fat and protein and fewer cases of mastitis and ketosis. In addition, breeding performance, in terms of conception rate and calving interval, is also improved.

DISADVANTAGES

Of course, all these benefits have to be weighed against such disadvantages as:

Cost

By comparison to most other means of providing supplementary feeds, inevitably capital investment requirements are higher, at around £65/cow per annum annual charge at 1990 prices. To this must be added the cost of concentrate storage and conveying. These investment costs may not necessarily be recouped by enhanced milk yields alone.

Cow refusal and training

It is not unusual to find a few animals which refuse to use feed stations. A certain amount of time may be needed to train cattle initially and to spend on new entrants to the herd. Even then there might be the odd cow which may continue to reject the proper use of feeders.

Equipment reliability

Although modern feeders are pretty reliable, some problems do occur often associated with faulty or lost transducers and feed dispensers can suffer from the consequences of working in a damp atmosphere. Ideally, any system should incorporate built-in alarms, not only for wholesale failure of a feed station but also if an individual cow is being deprived of her quota.

Limited potential for very high and very low yielding herds

It would probably be true to say that a herd producing 8000 litres from thrice daily milking, and already receiving three outside feeds, does not stand to gain a lot from out-of-parlour milking. Nor, presumably does a 5000 litre, 1 tonne concentrate/cow herd, since it hardly has much need to spread concentrate feeding.

Bullying

Behavioural studies show that timid animals may be dissuaded from making their due visits to feed stations by dominant cows. This is less of a problem where feed is split into several small meals with extended delay periods between. Bullying cows can also cause those lower down the 'dominance hierarchy' to vacate feeders prematurely. This tendency can be reduced by use of a suitably designed stall, preferably with either a lateral cross-bar or backing gate.

SPECIFIC APPLICATIONS FOR OUT-OF-PARLOUR FEED STATIONS

Grant and Eades (in 'Out-of-parlour feeding – is it for me?' *Dairy Farmer*, January 1984, 35–39) identified a number of circumstances in which out-of-parlour feeders might prove particularly appropriate. They are:

- Where large amounts of concentrates are fed in relatively few feeds of more than about 3–4 kg, and where cows could be bordering on clinical acidosis.
- Where more precision is required in individual feeding and physical grouping is difficult and/or time consuming.
- Where the calving pattern is rather spread, making fine tuning of concentrate feeding complicated both outside and in the parlour.
- Where the workload is high and time saved by automatic feeding can be profitably used on some other aspect of management.
- Where labour is in short supply and there is a desire to improve working conditions to attract good staff.
- Where the larger part of the herd is autumn calving, so that as much as possible of the total concentrates can be fed via feed stations.
- Where the existing concentrate feeding system is in need of major capital investment, for example new parlour feeders.
- Where a package of computerised equipment is planned to improve management.
- Where extra profit is not the prime motive and borrowing does not have to be justified on the basis of it.

AUTOMATIC OUT-OF-PARLOUR FEEDING STATION SYSTEMS

Nearly all are now based on cows wearing a transponder on a collar or in an ear tag. There are two main systems: one using an adjustable transponder and the other and more common one based on an auto recognition system.

Adjustable transponder system

The adjustable transponder system was developed at the University of Illinois in the mid-1970s. Although cheaper than auto recognition computer controlled systems, it does have the major disadvantage that each transponder needs to be individually adjusted on each cow when feed regimes have to be changed. Each transponder controls the amount of feed permitted to the cow over a 12-hour period. A cell in the transponder is charged at a controlled rate when the cow places its head in an electro-

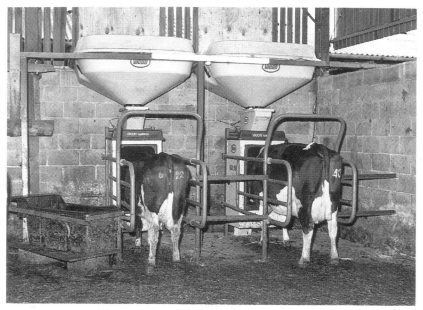

Plate 5.2 Out-of-parlour feed station (*Vicen*).

magnetic field created by an interrogator loop antenna at the manger. The transponder retransmits a signal which enables a feed dispensing auger to run as long as the transponder is being charged. When fully charged, the signal ceases and the auger is stopped. The transponder discharges at a linear rate and if not recharged would be effectively discharged after 12 hours. Frequent visits to the feeder by the cow will result in a smaller amount of feed per visit than if she visited only once or twice during the period.

Auto recognition system
Individual cow identity is provided by the signal emitted from a coded transponder worn by each cow. The device only transmits its coded information when energised by the interrogator at the manger. Typically, the controller can be pre-programmed for up to 255 cows. Having automatically checked whether that cow still has entitlement to any of its allocated ration it may then dispense a unit of feed to the manger and subtract this from the total ration allocation. The process is repeated after a relatively short time until the cow either vacates the feeder or has received its due amount for that period. The computer control unit is designed to spread the ration over a number of equal periods during 24 hours, so that only a proportion of the ration is available each period. The unit is automatically reset every 24 hours. Any of the previous day's feed allocation which has not been consumed can be brought to the stockperson's attention by

a 'morning alarm' which alerts the stockperson to any cows which have consumed less than a preset amount of feed, thus encouraging prompt attention to individual cows which may be sick. A regular print-out is also provided of the cows' intake against the programmed ration, giving a useful check on herd health.

Control over the amount and rate of feed dispensing varies from make to make. In one design, units of feed may be set from 0 and 1 kg and the daily amount dispensed per cow can range from 1 to 99 units. A trickle feed facility or a delay period of 1.5 minutes is organised between each unit of feed delivered providing, in effect, a delivery rate of 100–150 g/min, so that dispensing matches the eating speed of the cow and thus ensures that no cake is left for the wrong cow to finish. This feature also minimises bullying. A rolling memory ensures that a cow receives food at every visit to the feed station, but only in proportion to the time since her last visit. This is designed to ensure a regular intake of concentrate over a 24-hour period. In one design the day is divided up into 6-hour periods. In the first 6-hour period up to a quarter of the daily entitlement is available, a third of the remainder in the next 6-hour period, half the rest in the third period and the remainder in the last. These control features are typically regulated from a desk top computerised processor and display unit in the office or corner of the dairy.

Out-of-parlour feed stations comprise a master feed station which contains most of the electronics and one or more slave stations which merely take instructions to dispense feed from the master unit. Each feed station incorporates a race, or a feeding crate, to reduce the tendency for bullying, a feed manger, feed hopper and feed dispenser. Although some designs utilise a vibrating dispensing system, the majority employ a 12 or 24 V DC electrically driven auger system, which means that the unit can be battery powered for remote siting – say at pasture.

Among refinements now being offered are versions which enable two different types of concentrate to be fed in any combination from each feed station. The transponder signals to the computer to supply the required blend of feed for that particular cow. Feed delivery to each feed station is organised in one of three ways:

- By manual or batch loader filling of hoppers,
- By conveyor (usually centreless auger), or
- By gravity from a bulk bin.

For the last way a pair of feed stations can be sited under a lean-to shelter alongside a free standing bulk bin, periodically batch refilled by blower or auger. Installation should allow hoppers to be filled by hand in an emergency, and hoppers should be bird proof.

The maximum number of cows per station is related to access time and quantity of feed per cow. Usually the recommendation is 20–30 cows per feed station. ADAS suggest 250 kg divided by the average daily

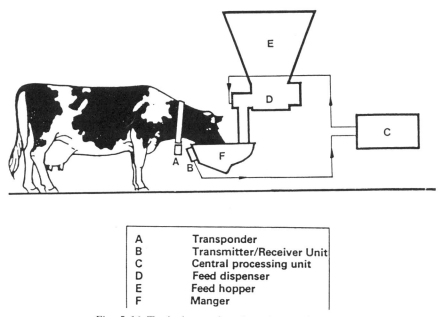

A	Transponder
B	Transmitter/Receiver Unit
C	Central processing unit
D	Feed dispenser
E	Feed hopper
F	Manger

Fig. 5.16 Typical out-of-parlour feeder (*Alfa Feed*).

concentrate intake (in kg/cow) should provide a reasonable estimate of the maximum. As animals are not forced to visit the station as they are with in-parlour feeders, it is important to site out-of-parlour feeders in areas where access is unhindered and is not liable to be blocked by antagonistic cows. The area round each feed station should be illuminated at night. Where the feed system is to be used during the grazing season, siting should allow easy access from pasture.

Out-of-parlour feeders require regular maintenance, maybe more so that in-parlour feeders as they may serve perhaps three times as many cows per day. The need for regular calibration cannot be over emphasised, otherwise data print-out for rations dispensed will be of limited value for management interpretation. A particularly useful feature of many types is the fact that the calibration figure can be entered into the control unit to adjust the quantity delivered for each unit and this avoids the need for mechanised adjustment of the dispenser itself.

Mechanised feeding of concentrates to beef cattle

Where concentrates are not mixed and fed with roughage, either by mixer wagon or conveyor mixer feeder then, for most beef cattle units, some form of mechanical feeding is normally worthwhile to minimise the chore of manual feeding. Whilst some barley beef units may use ad lib hoppers which can be filled, for instance, from reusable 'dumpy' bags transported by telescopic boom materials handler, most units use some form of restricted feeding in mangers. A number of possibilities exist, but much will depend on access limitations to buildings and mangers by machinery. Given adequate access to the type of tractor mounted, auger discharge feed boxes, as described in Chapter 2, may be appropriate for many situations. For flexibility it may be desirable for such a unit to be capable of discharge to either side.

For inaccessible strawed yards, concentrates can be automatically rationed and dispensed by using equipment similar to that used for feeding fattening pigs. Such a typical system comprises an overhead centreless auger conveyor supplying calibrated concentrate hoppers which, when tipped, discharge their contents into double sided free standing mangers beneath. Uniformity of distribution and hence consumption may be hard to achieve. Each hopper is capable of holding up to 40 kg concentrate feed.

A few beef units control individual access to different feeds for group housed animals by using 'key operated' access doors to specific feed troughs. Privileged animals wear a metal tie chain round the neck which, when it is brought near the electronic door lock, allows the animal access for feeding.

Mechanised liquid feeding of calves

Methods of liquid feeding

Figure 5.17 summarises the main calf liquid feeding systems. Any liquid calf feeding system must ensure that calves perform with a cost effective and acceptable rate of daily liveweight gain. Whilst for veal production the objective is to maximise consumption of liquid feed, for calves intended for dairy herd replacements and beef it is often more economical to restrict intake of milk substitute, introducing solid feed from a few days of age to stimulate rumen growth and so facilitate early weaning with the minimum of growth check. Good hygiene is a fundamental requirement with liquid feeding as scours can spread rapidly, particularly so with automatic group feeders.

For feeding of warm liquid milk substitute the choice is between automatic or semi automatic group feeders via teat outlets, often on an ad lib basis, or restricted feeding to individually penned calves, either by bucket or retractable teat bar. Though capital requirements are low with bucket feeding, it is a labour intensive choice and there are building layout limitations to provide access for individual feeding. In recent years, there has been growing interest in machine rearing, especially so for large dairy units where there is limited labour available for calf rearing. Automatic feeding group housed calves also provides a number of advantages for calf rearing, not least in flexibility of building use. However, cold liquid feeding is a less capital demanding alternative, very popular with many

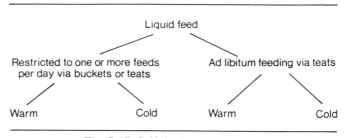

Fig. 5.17 Calf liquid feeding systems.

producers. On an ad lib basis calves suck proprietary cold liquid feed solution through teats up a plastic tube from buckets place outside the pen.

The major requirement for any mechanised liquid feeding system is to try to ensure that all calves are receiving sufficient, yet ideally restrict the intake per calf. Very high standards of calf management are required, including careful daily inspection of calves and close attention to hygiene aspects. Most compound manufacturers market special milk replacer powders for machine feeding, and some are linked with the sellers of the machine.

Mechanised aids to bucket feeding

Static feed mixers are a useful aid for preparing a thorough mix of milk substitute and water for either rationed bucket feeding or ad lib feeding from a storage container and teat bar system. Mixers comprise a steel or plastic container of capacity 80–160 litres, with a tap outlet, on a three-legged stand. Mixing is achieved by a motor driven agitating propeller in the base of the unit. Some mixers have the additional facility of a 3 kW thermostatically controlled immersion heater obviating the need for a separate supply of hot water.

For larger calf units it is possible to remove the chore of bucket carrying by distributing mixed feed from a suitable static mixing tank using an overhead insulated uPVC ring main with a series of strategically sited take-off points around the calf house. These comprise quick release bayonet type coupling points with on/off ball valves, into which a short section of delivery line is connected. Buckets are then filled to the correct level using a trigger operated nozzle outlet incorporating a probe and adjustable disc. The ring main length is limited to around 180 m to minimise the quantity of liquid contained in the system.

A mobile mixer/dispenser provides a flexible and less expensive alternative means of distributing feed for bucket feeding. Such a unit comprises a wheeled trolley with mixer tank of capacity about 225 litres and an electric pump for recirculation mixing and delivery of feed via a flexible plastic delivery pipe for filling of buckets using a similar trigger valve/probe and disc arrangement as used with the ring main system. The unit can also be used as a bucket washing plant, for dispensing water to each calf pen and for hosing down purposes when fitted with a restrictor on the outlet nozzle.

Semi automatic and automatic feeders

GENERAL CONSIDERATIONS

In general terms automatic calf feeders need to be justified on the basis of labour saving and convenience, rather than on the basis of improved performance and lower costs. However, group penning of calves does help in reducing calf housing costs. Ad lib feeders do provide means of maximising intake on a little and often basis – particularly important for veal production. But for dairy replacements and beef a number of authorities agree that the optimum policy should be to limit liveweight gain in the first three months to around two-thirds maximum level by restricted feeding of high quality milk replacer, fed in conjunction with ad lib hay and calf starter. Controlled restricted feeding of individual calves is not easy to achieve with most automatic group calf feeders. Extension of out-of-parlour feed control technology does now offer an answer to these problems.

A further consideration is that experience has shown a slightly higher level of calf mortality in automatic group rearing systems compared with single pen bucket systems, which reflects the difficulty of spotting when calves are ill. Group housed calves are more at risk to the possibility of rapid spread of infection and also to digestive upsets as a result of calves overindulging when first put on the machine.

TEAT BAR SYSTEM

This system uses a ring main layout similar to that used for bucket dispensing, but here the dispensing hose is connected to a teat bar along the front of individual pens of calves. Calves may have ad lib access, but more normally have restricted feeding. This is achieved by a lever mechanism, operated either manually or by a compressed air and timer system, to rotate the teat bar into the pens at feeding time. The system can also be used for group penned calves. A pressure regulator between the ring main and each teat ensures that the milk is at the correct pressure for easy drinking.

MILK/MILK SUBSTITUTE REHEATER-FEEDER

This is a popular low cost, semi automatic system of group feeding. Here, if over quota surplus milk is not used, milk substitute solution is mixed manually and stored in a suitable lidded plastic bin beside the reheater unit. This comprises a vessel of water, typically 30 litres in capacity, which is heated by a thermostatically controlled electric immersion heater. Milk

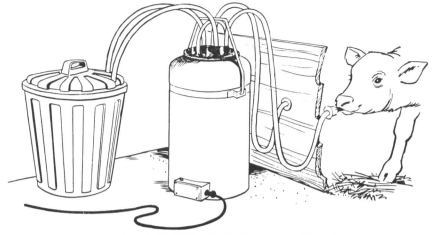

Fig. 5.18 Milk substitute warmer (*Volac*).

substitute, or milk, is drawn from the storage vessel through two plastic coils immersed in the warm water by the action of calves sucking on two teats. The solution is quickly warmed by heat transfer from the water reservoir to the plastic coil. The water heating thermostat is normally set to around 45°C to allow the solution to be available at the teats at around 38°C. Teat lines should not exceed 1 m in length to avoid excessive heat loss. The milk/milk substitute storage vessel needs to be cleaned daily and replenished at regular intervals.

AUTOMATIC GROUP CALF FEEDER

Most automatic group calf feeders provide ad lib access to liquid feed, although some incorporate the facility to limit intake by means of built-in delay periods between feeds. A typical feeder unit has a capacity of up to 50 calves in a maximum of four separate groups, each provided with a teat outlet and non-return valve supplied from the feeder through a plastic tube. The unit comprises a milk substitute powder hopper of capacity 33 kg, a 25 litre water tank replenished through a float valve, water heater, feed dispensing auger, water pump, mixing vessel with agitator paddle and a control panel. Water in the supply tank is maintained at a near constant temperature by a 2 kW water heater and thermostat. When a level probe in the mixing vessel is no longer in contact with liquid the control system initiates a mixing cycle. For each half litre of warm water pumped into the mixing vessel around 50 g milk substitute powder will be augered in, depending on the desired concentration rate. Once the level probe is covered another cycle of water and powder is delivered to give a maximum reservoir supply of up to 2 litres. Ingredients are mixed by the beater paddle as they are being added and for a further 15 seconds

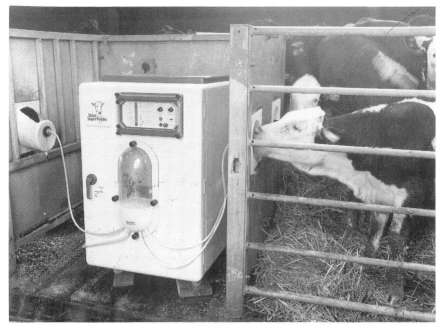

Plate 5.3 Automatic group calf feeder (*Volac*).

afterwards. The mixing cycle will then stop and hold until the mixed liquid is drawn off to just below the level probe. To ensure adequate dispersion when there are no regular mixing cycles any solution not consumed is agitated every 120 seconds. Mixed solution is kept up to temperature by heat transfer simply because the mixing vessel is moulded into a recess in the hot water tank. Tube length to each pen must not be more than about 2 m to minimise hygiene and cooling problems.

Solution strength is adjustable in the range 8–18% solids, with 10–14% being the most usual. A solution strength of 8–9% is recommended for the first few days that calves are introduced to the machine.

The control system 'fails safe' if the level of solution does not reach the bottom of the probe within 90 seconds, for instance if the door had not been screwed down tight, after cleaning, and was leaking.

Group feeder management

(1) CLEANING AND CALIBRATION

The hopper is topped up daily and the mixing bowl and milk lines should be drained on a daily basis as well, and washed out with hot water and a suitable cleansing agent. At the same time it is recommended to carry out a simple calibration check by collecting 1 unit of water in a measuring

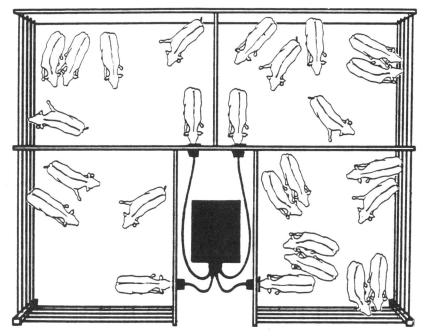

Fig. 5.19 Typical layout for group segregated feeding of dairy herd replacement calves.

jug and then 1 unit of milk powder and weighing it. Any adjustments necessary are achieved by turning either the water or powder control knobs which control the duration of the running of the pump and dispensing auger.

(2) MILK INTAKE
If milk intake is considered to be excessive for the chosen overall rearing system some reduction can be achieved by decreasing the solution temperature. Maximum intake occurs when the feeding temperature is 38°C. A reduction of temperature to 30°C can result in a 20–50% decrease in milk substitute intake. Though it is not a normal recommendation, individual calf intake can also be limited by increasing group size, and there is some evidence that this can be achieved without a significant deterioration in group variation in weight gain.

(3) CALF GROUPS
It is desirable that calf groups should be of similar age and weight for the best results. When groups are segregated into four groups, as Fig. 5.19, to accommodate a regular intake of calves, each group at a different age and weight, then ten calves per group is the optimum. No matter what

the group size, it is desirable that no more than twelve calves should be allocated per teat. It is usual for teats to be replaced every batch.

(4) CALF TRAINING AND OBSERVATION
In order to identify non-drinkers and encourage suckling in the first 3–4 days, calves should be disturbed and made to stand every 4 hours or so. A close and regular eye needs to be kept on calves twice daily throughout the ad lib feeding period to spot any calves which are 'off colour'. The requirements for high standards of stockmanship cannot be over emphasised.

(5) MILK SUBSTITUTE SUPPLY
It is vital that feed flows constantly from the machine's hopper into the mixing vessel. Thus, milk substitute powder should be kept dry and free from lumps. If required, a vibrator unit can be fitted to the side of the powder hopper.

(6) ACCESS TO WATER, HAY AND CALF STARTER
Clean water should be available from three weeks, together with a supply of hay. Calf starter pencils should be available from the second week.

(7) BEDDING
Ad lib fed calves produce larger quantities of urine and it is essential that there is not only adequate dry bedding, but also appropriate fall on the floor and suitable drainage.

(8) WEANING
If feed consumption is to be kept at a realistic level early weaning must be achieved. This can be achieved in a number of ways. Thus, the feeder can be turned off for regular periods or the tests for appropriate pens unscrewed and removed for specific periods. Alternatively, the solution temperature may be reduced, the flow reduced by fitting restrictors in the supply tubes or the tubes removed from the machine and the ends placed in a bucket of water.

Computerised liquid calf feeders

For those dairy producers already operating a computerised out-of-parlour feeding system it is a comparatively easy step to add a worthwhile refinement to automate fully the calf group feeding system. Calves are fitted with individual neck bands and transponders which facilitate close control on the quantity and frequency of feeds according to individual calf needs. But the most significant feature is that the computer program promptly 'flags up' those calves which are not receiving their due allowances, thus alerting the stockperson to take early remedial action. Each teat outlet is incorporated into a feed station with interrogating antennae and metal or wooden stallwork to minimise the likelihood of the calf being pushed away by others and also to position the calf correctly and ensure proper interrogation by the control system.

Early reports indicate that very significant savings of milk substitute powder are being achieved. With milk substitute powder now costing over £1000/tonne and perhaps an average consumption of 1 kg or so per day over a 5-week weaning period, there are promising signs that transponder feeding of calves will become progressively more popular.

References

ADAS (1983) *Out of Parlour Feed Dispensers*. Ministry of Agriculture, Fisheries and Food. Publication DHMP/24.

ATI (1988) *Handbook of Milking Parlour Feeding and Control Equipment*. ATI Agricultural Technology Ltd.

Buck, N.L. et al (1987) 'Performance of electronic animal identification in the milking parlour, *Journal of the American Society of Agricultural Engineers* **3**(2), 153–158.

Chancellor, G. (1984) 'Out of parlour feeding – is it for me?' *Dairy Farmer* January, 34–39.

Dawson, J.R. (1980) *Equipment for Feeding Concentrates in and out of the Parlour. The Mechanisation and Automation of Cattle Production*. British Society of Animal Production Occasional Publication No 2, 155–169.

Gibbs, M. (1988) 'Justified – out of parlour feeding'. *Dairy Farmer* June, 16–18.

Kilkenny, J.B. (1980) *Mechanisation of Liquid Feeding of Calves*. British Society of Animal Production Occasional Publication No 2, 115–124.

Leaver, J.D. (1987) 'Importance of feed costs.' *Farm Buildings Association Winter Conference Report*, 1–4.

MAFF (1975) *Mechanised Parlour Feed Dispensers*. Mech Leaflet No 34.

Marshall, I. (1983) 'Out of parlour feeders.' *Power Farming* June, 16–19.

McCarthy, T.T. (1987) 'Electronics, identification and control.' *Farm Buildings Association Winter Conference Report*, 14–18.

MMB (1983) *Computerised Out of Parlour Feeders*. FMS Report No 38.

Scottish Agricultural Colleges (1987) *In Parlour Feeder Survey*. Technical Note No 108. West of Scotland Agricultural College.

Taylor, S. and Green, D. (1988) 'Volac'. Personal communication.

Thickett, W., Michell, D. and Hallows, B. (1986) *Calf Rearing*. Farming Press, Ipswich.

Turner, M.J.B. (1984) 'Concentrate feeding'. *Agricultural Engineer* **39**(4), 141.

6   Mechanised poultry feeding

All large poultry units are virtually totally dependent upon efficient and reliable mechanised systems of feeding birds. Such systems require careful planning to ensure continuity of feed supply and adequate access, not only by birds in feeding but also by stockpersons for maintenance and repair work. At cleaning out time in littered houses it may be important that feeding equipment can be shifted quickly – for instance by winching up into the roof space. Many poultry producers agree that, after reliability, reduction of feed wastage should have next priority as birds have the wasteful tendency to flick feed from the trough as they eat. Close attention to trough design and depth of feed, plus height of trough above the floor, will have significant effects upon feed wastage. Some birds are more aggressive than others in their feeding habits, for example turkeys, and trough design refinements, such as special lips to retain flicked feed, are particularly worthwhile. All mechanised poultry feeding systems certainly justify the incorporation of some form of accurate monitoring of feeding consumption, for instance by tipping weigher or load cell weighing. Not least, attention should be paid to having some form of back-up feeding system available in the event of mechanical breakdown – and this requirement is now encompassed within the Welfare Codes.

System design factors

(1) Type and quantities of feed

Feed is made available as meal, pellets or crumbs, the last two forms being less prone to dust which, with meal, tends to be dispersed by birds as they flick the feed with their beaks. But there is the tendency for pellets to be consumed too quickly by cage housed birds and empty feed troughs in front of the birds can give rise to vices such as feather pecking.

It is usual for feed to be stored in outside free standing storage bins – which should be of sufficient size to provide adequate back up stocks – even when the birds are at peak consumption. Alternatively, feed may be transferred directly from an adjacent on farm feed preparation unit.

Depending upon the size and type of hybrid – feed will be fed at a maximum rate of 85–140 g/day for laying birds and up to 170 g/day for table chickens. Details of bulk storage bins are given in Chapter 2.

(2) Distribution of feed

Whatever conveying system is selected to distribute feed to troughs or pan feeders there should be no tendency for separation of coarser and finer materials nor should there be any inadvertent grinding, which will only aggravate the dust and waste problem. Feed should also be distributed uniformly throughout the house. With many systems it is usual to transfer feed automatically from an outside storage bin to a supply hopper inside, from which feed is then conveyed to troughs or feeders. A proximity diaphragm, or flap switch, in the supply hopper automatically controls the level of feed delivered from the storage bin. In most cases it is advisable for the supply hopper to be located in a separate room adjacent to the main poultry floor, so that filling and maintenance can be done without disturbing the flock, and also to keep the birds off it.

(3) Space requirements

Feeding space requirement for chain feeders will depend upon size of bird and whether ad lib or restricted feeding is practised. As a guide, each linear metre of trough with access both sides will cater for 50–80 broilers and 16–25 birds. Because of the way that the birds stand around the feeding place, pan feeders give about one-third more feeder space than the same length of straight trough.

(4) Reliability

The poultry producer must have confidence that equipment will have a high sustained reliability factor and it obviously pays to pick a reputable make with an assured 24-hour repair back-up service. As with much equipment, the more moving parts the more maintenance needs and likelihood of breakdown, so there is a tendency to keep the system relatively simple. Adjustments should be straightforward and repair of chain or cable conveyors should be capable of being carried out quickly and with the minimum of skill. Relevant spare parts must be kept to hand.

The system should always 'fail safe' so that, for instance, supply augers do not overfill hoppers and spillage on return to the feed hopper is avoided, and it may be very worthwhile installing a feed alarm system. This gives an 'alarm' output to a bell/hooter/flashing light if the feed chain breaks, the motor drive fails or the shear pin breaks.

As well as a standby generator, an emergency manaul back-up feeding system should be available when repairs cannot be effected within, say, half a day, and the producer must be mindful of his or her obligation to make such provison under the Welfare Codes.

(5) Monitoring feed quantities

Outside storage bins need some means of identifying the quantity of material left in them as they are being emptied. This can be achieved by simply using a transparent panel in the bin which will, however, tend to be clouded by dust, or by using a probe switch mounted in the side of the hopper corresponding to, say, a third of the capacity remaining. At the appropriate point the probe switch will operate a warning bell or lamp. Where home mill and mixing is practised, a similar probe switch arrangement can be used to instigate the automatic transfer of rations from the mill and mix plant to replenish storage bins.

However, the discerning poultry producer really does need on-going, reasonably accurate information on quantities of feed being consumed. This can be achieved in a number of ways. A particularly convenient method is to mount each storage bin on strain gauged load cells, which will also obviate the need for a separate probe switch arrangement. Alternatively, particularly where feed may be transferred to houses directly from an adjacent on-farm mill and mix unit, a tipping weigher can log the amount of feed transferred to a particular house.

A number of poultry feeding systems do incorporate a weighing facility associated with the internal supply hopper which, in one system, is suspended from the ceiling by a spring balance. In another system, the supply hopper is mounted on platform scales. In each case manual logging is required every time the hopper is replenished, but an automatic logging facility can be organised with electronic weighing systems.

(6) Adjustment of feeders

For floor/slat fed birds, it is important that the depth of feed in the trough or pan feeder is adjustable in order to minimise waste. As birds grow the feeders will need to be raised so that they are kept at about the height of the fowl's back. This is effected by adjustable support legs for chain feeders, or by a cable winching arrangement or a combination of both. For winching it is worth ensuring that the roof members are capable of taking the extra imposed loads of fully laden conveyor feeders.

Where pan feeders are supplied by low level tube conveyors it is usual to fit a spring tensioned wire a few centimetres above the top of the conveyor. Usually this wire is electrified in a similar way to an electric fence for cattle. This arrangement is intended to dissuade birds from perching on the conveyors and hence severely loading the roof members.

Cable winching can be effected manually by winding in the cable on a capstan, or it can be fully mechanised using a motor/gearbox or linear actuator, and this can be fully automated by means of a simple programmer. One novel semi automatic system provides for a standard electric drill to be coupled up to power the drive to the winching arrangement when required. One particular advantage of cable winching, by contrast to an adjustable leg system, is that feeders and supply conveyors can be conveniently winched into the roof space, clearing the floor area for catching birds and cleaning out.

(7) Food contamination

One particular problem with chain open topped trough feeders is that droppings and other foreign matter can accumulate and contaminate feed. To cope with this a food cleaner is fitted to the end of the circuit before the chain enters the hopper for refilling. The most common type comprises a rotating screen which is driven by the movement of the chain. Feed is diverted from the chain into the centre of a vertically rotating screen. Fine particles drop through the mesh and are then lifted back on to the chain conveyor, while large particles are 'screwed' outwards and dropped on the floor. Appropriate sized screens are needed for meal, crumbs or pellets.

(8) Feeding programmes

Whilst some systems operate on an ad lib basis, it is usually more efficient to incorporate some form of restricted feeding, which should decrease food wastage and allow time between meals for more efficient digestion. Furthermore, it is often observed that intermittent feeding encourages appetite. Caution does need to be exercised, however, as feed efficiency may be significantly decreased if broilers are without feed for more than one hour between feeding cycles. Adequate feeding space and rapid feed

Plate 6.1 Rotary screen feed cleaner (*Big Dutchman*).

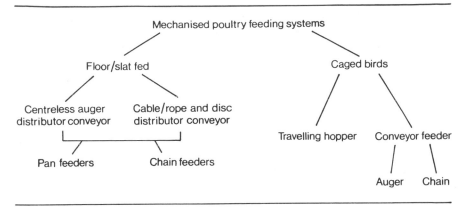

Fig. 6.1 Mechanised poultry feeding systems.

distribution are essential for this type of management practice. For caged birds a further problem with restricted feeding is that if food is withheld for more than half an hour or so, then boredom encourages vices such as feather pecking.

Thus, feeding is normally programmed, using timeclock control, to provide between four and six separate feeds during the day. One pan feeder manufacturer recommends that for broilers the feeding day should be organised into four periods of 6 hours, with the aim to have the pans empty for a period up to one hour at the end of each 6-hour period. Alternatively, for chain trough feeders, boredom can be avoided and consumption stimulated by a programme comprising feeding for 1 hour, stopping for 90 minutes, then feeding for 15 minutes and repeating the 90 minutes off, 15 minutes on cycle until the next 1-hour feeding period is reached the following day.

For layers in cages it is usual to feed five or six times per day, but the birds should have ample time to feed before 'lights out'.

Mechanised poultry feeding systems

Figure 6.1 summarises the most common mechanised systems of feeding poultry.

Floor/slat feeding systems

CHAIN FEEDERS
The system comprises a supply hopper with drive unit, one or more complete circuits of feed trough mounted on adjustable legs, conveyor chain, special corner units, rotary screen feed cleaner and control system.

Depending on the number of circuits to be served the supply hopper, sited in a room adjacent to the birds, will hold 70–120 kg feed, but this can be increased to 600–700 kg by the use of extension sides. This hopper has one vertical side and one 60° slope to ensure rapid abstraction of material by the chain which runs through the bottom of the hopper, driven by a motor (of 350–800 kW, depending on chain speed and length) through a gearbox and sprocket drive. The motor may also drive an agitator to assist in meal movement. The chain, comprising flat, flexible interconnecting links, draws feed along in the bottom of an open topped trough of approximately 75–85 mm width at the base with varying shapes and widths at the top for different types and weights of birds. Special links are used for connecting sections of trough to form a continuous circuit. The links may be fitted to support leg assemblies for vertical adjustment at the desired level. Special horizontal corner sections, again mounted on adjustable legs, enable the trough circuit to negotiate 90°, 135° or even 180° bends, with the aid of a pulley wheel arrangement. Chain is tensioned with the aid of a tension unit comprising an adjustable telescopic section of trough, which can be extended by a screw thread.

The depth of feed is controlled by a sliding guillotine in the supply hopper and feed is distributed at a chain speed of 6–18 m/min. Where feed is restricted it is important that birds feed at the same time and thus a chain speed of at least 12 m/min is needed to ensure that feed is distributed quickly to all birds within a short period of time. It is recommended that the feed chain should make a complete circuit within 10 minutes, and should run until the allocated feed has been consumed. Thus, if the total travel is more than 120 m the system should be divided up into two or more circuits, which will probably be necessary anyway in order to provide sufficient trough space for the number of birds housed. Thus, for a 12 m wide house the general rule is for a double layout giving four rows of troughs down the length of the house, with appropriate height adjustment to allow birds to eat comfortably.

Multicircuit units call for the supply hopper to accommodate two or more chains passing through to pick up feed. The hopper will also need to be fitted with an extension to ensure that there is sufficient supply of feed available.

A rotary screen feed cleaner is fitted just before the chain enters the hopper for refilling, to remove droppings and other foreign matter from the feed and thus reduce waste.

By the nature of its design, a chain feeder does cause feed to be carried with variable consistency and as it scrubs the trough sides it causes a certain amount of feed separation. An improved chain feeder developed at the National Institute of Poultry Husbandry is not only reversible but incorporates special plastic links with vertical sections which, it is claimed, ensure that feed is conveyed more efficiently, enabling more feed to be placed at a time and with less feed separation.

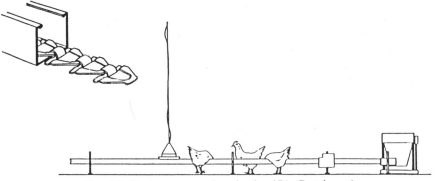

Fig. 6.2 Chain type poultry feeder (*Big Dutchman*).

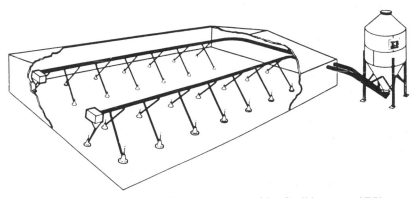

Fig. 6.3 Typical pan feeder system served by flexible auger (*EB*).

An alternative design of chain feeder is now imported from the USA. It comprises a chain of double circular discs linked by short metal rods and which, it is claimed, has only one-third the weight per metre of flat chain, enabling faster chain speeds of up to 30 m/min and a smaller motor. The chain runs in the bottom half rounded section of a V shaped trough. As well as resulting in faster and more efficient distribution, it is claimed that the trough design does help to minimise feed wastage.

PAN FEEDERS
Pan feeders offer a number of advantages over chain type trough feeders. The latter do create a barrier across the house, restricting the areas in which the birds may move, whereas pan feeders increase the ability of birds to move over the whole floor area. Also, as stated earlier, pan feeders offer around one-third more feeder space than the same length of trough, because of the way birds stand around the feeding place. There is a certain

amount of evidence, too, that feed wastage and contamination are less than with trough feeders.

The system comprises a series of circular floor mounted or suspended pan feeders supplied by either cable and disc tube conveyor or centreless auger conveyor. There are many varieties on pan feeder design and most are made from moulded plastic. Some comprise two parts only: a circular pan with inward curving lip and a cone to receive feed and allow it to be dispensed downwards around its bottom periphery. Other pan feeders incorporate a slatted grille to segregate birds and also to help minimise waste.

Depth of feed can be adjusted by raising or lowering the central cone on pre-set holes on a central rod with locating pin. Alternatively, on some pan feeders the control cone can be twisted and raised up or down on a series of stepped notches. Whilst most pan feeders are very simple in design and are applicable for ad lib or restricted ad lib feeding, one particular version is available with electric auto cut-off which provides the facility for restricted feeding.

Filling of pans is achieved either from high or low mounted centreless auger conveyors with drop tubes to feeders or from a low level circuit of cable and disc conveyor.

Centreless auger feed distribution system

This is a simple system suitable for a relatively limited number of lines of feeders. Figure 6.3 shows a typical recommended layout. Although more than one drive unit may be needed, the main advantage over most cable

Plate 6.2 Typical pan feeder (*Chore-Time*).

and disc conveyors is that no separate internal supply hopper is needed, nor any separate room to house it, and bends of around 3 m can be accommodated without the need for corner units. A typical system would also have a much greater carrying capacity. In many installations with high mounting, no anti perching wires are needed, nor any special winching arrangements. For house cleaning purposes, drop tubes pivot about swinging T-pieces on the main auger tube, and can be swung up and secured in the roof space. It is usual for each auger tube to supply two staggered rows of pan feeders, although it may be possible to accommodate three rows. It is recommended, however, that drop tubes should be inclined at an angle of at least 60° to the horizontal to ensure unimpeded flow of feed. Maximum permitted auger length is around 120 m.

Cable and disc feed distribution system

Cable and disc conveyors are not only quieter running than auger conveyors but also accommodate a complete, and perhaps tortuous, circuit of up to 500 m needing only a single drive unit, using conveyors of 35–60 mm in diameter. Details of these conveyors are further discussed in Chapter 2. It is usual for pan feeders to be directly attached to the conveyor tube and suspended above the floor by a cable winch system.

The conveying system comprises loading hopper, drive unit, endless cable with disc attached and special corner units. The loading hopper is normally replenished by auger from an outside bulk bin. If required, a vibrating sieve can be fitted to remove foreign bodies from the feed. Feed is then fed on to an exposed section of conveyor in the loading hopper by an auger driven by a spiked wheel which meshes with, and is rotated by, the conveyor travel. The drive unit comprises a motor and gearbox drive with double or treble sprocket system to effect cable drive and manual and automatic tensioning is achieved by spring loading on an idler sprocket. An electronic rotation control system can be fitted which will immediately stop the system if it becomes overloaded. The drive unit also incorporates a spring loaded switch arrangement that switches off the motor should the cable break.

Discs are made of special plastic, which is resilient yet durable, and the discs are bonded on to steel cable or nylon rope. It is preferable that the cable joining system is simple enough to avoid the need for special tools. Corner units comprise idler sprockets mounted in either pressed steel or, better, cast aluminium alloy shells, with the sprocket ideally mounted on anti friction bushes or ball bearings. It is desirable that, especially for outside use, a sealant is incorporated into the two exposed faces of the casing before they are bolted together. In one version, a special corner unit is available for fitting before the cable returns to the drive unit. This corner unit incorporates a plastic toothed sprocket which effectively 'lines up' the discs ensuring good disc and sprocket register and thus

Plate 6.3 Cable drive unit (*AZA*).

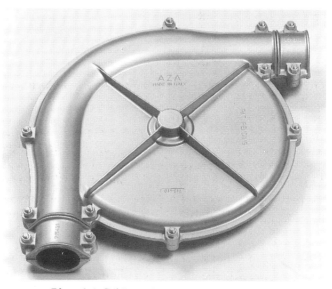

Plate 6.4 Cable and disc corner unit (*AZA*).

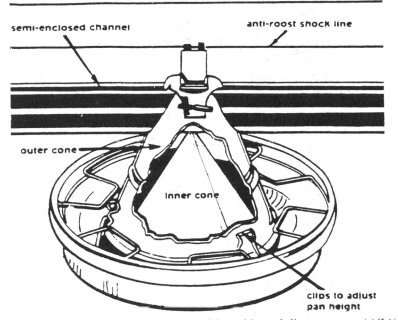

semi-enclosed channel

anti-roost shock line

outer cone

Inner cone

clips to adjust
pan height

Fig. 6.4 Typical pan feeder system served by cable and disc conveyor (*AZA*).

minimising disc/sprocket damage and wear. However, in one version discs on the cable do not need to fit specifically into notches on the drive sprocket and therefore it is not necessary to have exactly the same intervals in the connecting link as in the rest of the cable.

Mechanised feeding of caged birds

There are two basic systems used: travelling hoppers and conveyor feeders.

TRAVELLING HOPPERS

With travelling hoppers a series of tapered hoppers, one per layer of cages, is filled up at one end of the row of cages by an overhead auger. There is a need for a flap or lever operated switch to stop delivery when the last of the hoppers is completely filled. Similarly, a limit switch may be desirable to prevent augers restarting when hoppers move away to fill the cage troughs. Hoppers are moved along slowly by cable winch and feed distributed to a depth set on a feed gate at the bottom of each hopper. When the hopper unit reaches the far end a reversing switch effects automatic return of the unit for refilling.

Although the length of the cage row is limited by the hopper holding capacity, hopper feeders do have certain advantages in that capital cost

Plate 6.5 Typical chain type feeder for battery cages.

is lower than chain feeders and, if need be, they can be hand filled and manually winched or pushed along for emergency feeding purposes.

CONVEYOR FEEDERS

Conveyor feeding of caged birds does require less space and the conveyor action is said to stimulate the birds to eat. There are two main systems in use: chain feeders and an auger system.

Chain feeders

Chain feeders are of two common types, using either flat or round chain. Each back-to-back layer of cages is served by a continuous loop of chain running in the base of the feed trough served by a sprocket wheel drive unit. For flat chain systems, drive from a top mounted motor and gearbox unit is transferred by a sprocket and chain drive transfer unit to chain conveyors serving each layer of cages. Round chain systems use a separate

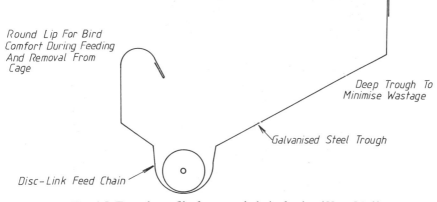

Round Lip For Bird
Comfort During Feeding
And Removal From
Cage

Deep Trough To
Minimise Wastage

Galvanised Steel Trough

Disc-Link Feed Chain

Fig. 6.5 Trough profile for round chain feeder (*Hart-Link*).

drive unit for each layer. Feed is supplied to the chain at the end of each layer of cages via a multi tiered feed intake hopper unit which is replenished in turn by auger or cable and disc conveyor. The depth of feed in the trough is controlled by an adjustable guillotine gate arrangement.

Auger feeders
In auger feeders, a centreless auger transfers feed from the supply hopper around the troughs in a complete circuit of the block of cages, exploiting the auger's considerable flexibility in accommodating changes of direction both horizontally and vertically. Birds cannot feed until the auger has completed dispension of feed and it is claimed that the auger dimensions provide segmented eating places, which dictate that the bird has to eat in a controlled way thus minimising the amount of feed which can be flicked out of the trough.

In all battery cage feeding systems, potential feed wastage is significantly affected by the design of the feed trough profile and the depth of feed maintained in the trough. In general terms, the trough needs well designed lips which should be higher on the outside edge to minimise the amount of feed which can be 'billed' out on the floor, and the depth of feed should be minimised.

References

Electricity Council. (1974) *Automatic Feeding of Poultry and Pigs*. Farm Electric Handbook No 20.

Forster, M. (1985) 'Feeding systems for broilers.' *Gleadthorpe EHF Poultry Booklet*, 4–10.

Lambert, I. (1987) 'Feeding systems for pigs and poultry.' *Farm Buildings Association Winter Conference Report*, 23–27.

Skipper, M. (1988) 'TFS'. Personal Communication.

Index